Developments in Petroleum Science, 6

FUNDAMENTALS OF NUMERICAL RESERVOIR SIMULATION

FURTHER TITLES IN THIS SERIES

  1 A.GENE COLLINS
GEOCHEMISTRY OF OILFIELD WATERS

  2 W.H. FERTL
ABNORMAL FORMATION PRESSURES

  3 A.P. SZILAS
PRODUCTION AND TRANSPORT OF OIL AND GAS

  4 C.E.B. CONYBEARE
GEOMORPHOLOGY OF OIL AND GAS FIELDS
IN SANDSTONE BODIES

  5 T.F. YEN and G.V. CHILINGARIAN (Editors)
OIL SHALE

Developments in Petroleum Science, 6

FUNDAMENTALS OF NUMERICAL RESERVOIR SIMULATION

DONALD W. PEACEMAN

Senior Research Advisor,
Exxon Production Research Company,
Houston, Texas, U.S.A.

ELSEVIER SCIENTIFIC PUBLISHING COMPANY
Amsterdam — Oxford — New York   1977

ELSEVIER SCIENTIFIC PUBLISHING COMPANY
335 Jan van Galenstraat
P.O. Box 211, Amsterdam, The Netherlands

Distributors for the United States and Canada:

ELSEVIER NORTH-HOLLAND INC.
52, Vanderbilt Avenue
New York, N.Y. 10017

Library of Congress Cataloging in Publication Data

Peaceman, Donald W
  Fundamentals of numerical reservoir simulation.

  (Developments in petroleum science ; 6)
  Bibliography: p.
  Includes index.
  1. Oil reservoir engineering--Mathematical models.
2. Oil reservoir engineering--Data processing.
I. Title. II. Series.
TN871.P37   622'.18'282   77-4771
ISBN 0-444-41578-5

ISBN: 0-444-41578-5 (vol.6)

ISBN: 0-444-41625-0 (series)

© Elsevier Scientific Publishing Company, 1977.
All rights reserved. No part of this publication may be reproduced, stored in a retrieval system or transmitted in any form or by any means, electronic, mechanical, photocopying, recording or otherwise, without the prior written permission of the publisher, **Elsevier Scientific Publishing Company, P.O. Box 330, Amsterdam, The Netherlands**

Printed in The Netherlands

PREFACE

Over the past decade, the use of numerical reservoir simulation with high-speed electronic computers has gained wide acceptance throughout the petroleum industry for making engineering studies of a wide variety of oil and gas reservoirs throughout the world. These reservoir simulators have been designed for use by reservoir engineers who may possess little or no background in the numerical mathematics upon which they are based. Yet in spite of our best efforts to improve numerical methods so as to make reservoir simulators as reliable, efficient, and automatic as possible, the user of a simulator is constantly faced with a myriad of decisions that have nothing to do with the problem he really wants to solve. He must decide on various numerical questions not directly germane to the problem at hand. For example, he may have a choice among several simulators that use different numerical methods. He may have to pick an iteration method. He definitely will have to choose the grid spacing as part of the reservoir description, and probably will also have to select the time step size. And perhaps the biggest bugaboo of all is the choice of iteration parameters.

It is this engineer-user that I have had in mind while writing this book, one who wants to learn how to deal more effectively with the numerical decisions mentioned above. I hope he also has some curiosity about the inner workings of the "black box" that is a reservoir simulator, and I have tried to satisfy that curiosity, as well as to prepare him to read the literature, should he wish to study recent developments and future research in greater depth than I have been able to provide here.

The first chapter combines a review of some basic reservoir mechanics with the derivation of the differential equations that reservoir simulators are designed to solve. The next four chapters provide basic theory on the numerical solution of simple partial differential equations. The final chapter brings together this basic theory as it applies to the numerical solution of multidimensional, multiphase flow problems.

I have attempted to make this book as self-contained as possible. The reader is assumed to have some knowledge of partial differential equations and simple matrix algebra; additional mathematical tools are provided where needed. In developing the numerical theory, I have tried to serve the engineer's needs better than do the standard textbooks on numerical analysis, which tend to be either too rigorous or too general. I have not attempted to be completely rigorous in the mathematical proofs, but I have included sufficient derivations so as to make the various mathematical arguments as plausible as possible.

But the engineer-user is not the only reader I had in mind in writing this book. The mathematician skilled in numerical analysis will, of course, find much material already familiar to him. However, the first chapter will introduce him to the basic principles of reservoir mechanics, and the remainder of the book will indicate to him those topics in numerical analysis that I consider significant in the numerical solution of reservoir flow problems. Furthermore, I have included some material which appears not to be well known. For example, the section on successive overrelaxation methods discusses the effect of Neumann boundary conditions and the effect of anisotropy, whereas standard textbooks confine themselves to Dirichlet boundary conditions and isotropic problems. Finally, the material in the last chapter should be new to most numerical analysts, as it is quite special to the area of multiphase reservoir flow. It is my hope this book will provide food for thought leading to further progress in numerical reservoir simulation.

While much work is now being done on the application of variational methods to the solution of partial differential equations, little of this has reached the stage of practical application in reservoir simulation. Most practical reservoir simulators now in use are based on finite-difference methods. For this reason, only finite-difference methods are covered in this book.

The reader familiar with reservoir engineering will note a departure from a practice I feel is all too common in the field, namely, the inclusion of numerical constants in equations involving flow. All the equations in this book are free from numerical constants (which are dependent on the units being used) and are valid for any consistent set of units. The use of dimension-free equations should become more common as the industry moves to the adoption of the SI (Système International) standard of units, as is now being proposed. Accordingly, in the nomenclature following each chapter, I have specified the units of various quantities in the basic SI units of kilograms, meters, and seconds, together with the derived units of the newton for force (which equals $kg \cdot m/s^2$) and the pascal for pressure (which equals N/m^2). These form a consistent set of units. If the reader prefers, any other consistent set of units can be used, and the equations will still be correct.

The material in this book is based primarily on notes prepared for a series of lectures I was privileged to give at a NATO-sponsored Summer School on "Hydrocarbon Reservoir Simulation by Computer", held in Milan, Italy, in May, 1969. I have given the same lectures within Exxon Production Research Company and for a Continuing Education Group of the Los Angeles Basin SPE Section in January, 1974. Many who have seen these lecture notes have urged me to enlarge and publish them. I felt that the original notes were somewhat incomplete and so added several sections. The most significant additions were the section (in Chapter 1) on alternative differential equations for two-phase flow, all of Chapter 4 on solution of hyperbolic problems, the section (in Chapter 5) on relaxation methods, and the sections (in Chapter 6) on sequential solution methods, semi-implicit mobility, and well rates.

I am indebted to the management of Exxon Production Research Company for permission to publish this book and for the encouragement they gave. In particular, I want to thank C.C. Mattax, Manager of the Reservoir Divison, for his encouragement and many helpful suggestions about the revision of the original lecture notes. Thanks are due, also, to J.W. Watts and R.P. Kendall, of EPR, for the helpful discussions I had with them while writing the section on relaxation methods. Discussions with J.G. Hillestad and H.L. Stone, of EPR, also were of great help.

I owe much to the skill of Dodi Fenner for the preparation of the figures, to Winn Alms for typing the original lecture notes and part of the book manuscript, and to Altha Frazier and the Word Processing Group of EPR for typing the major portion of the manuscript for the book. I owe a large debt of gratitude to John Colby, Supervising Editor at EPR, for carefully editing the first draft of the book and checking the printer's proofs.

Finally, I would like to dedicate this book to the three women in my life, who have provided inspiration to me and many others around them: to my mother, Ida, of blessed memory; to my wife, Ruth; and to my daughter, Caren.

DONALD W. PEACEMAN

Exxon Production Research Company,
Houston, Texas,
January, 1977

CONTENTS

PREFACE V

Chapter 1. DIFFERENTIAL EQUATIONS FOR FLOW IN RESERVOIRS 1

Introduction 1
Single-phase flow 2
   Darcy's law 2
   One-dimensional, single-phase, compressible flow 3
   Two-dimensional, single-phase, compressible flow 5
   Three-dimensional, single-phase, compressible flow 6
   Differential operators 6
   General equation for single-phase flow 7
   Boundary conditions 8
   Special cases of single-phase flow 9
      Ideal liquid of constant compressibility 9
      Liquids of slight compressibility 10
      Ideal gas 11
      Incompressible flow 12
Types of second-order differential equations 12
   Parabolic equations 12
   Elliptic equations 13
   Hyperbolic equations 13
   Classification of equations 13
   First-order hyperbolic equations 13
Two-phase flow 14
   Introduction 14
   Darcy's law 15
   Conservation of each phase 16
   The differential equations for two-phase flow 16
   Alternative differential equations for two-phase flow 17
      Pressure differential equation 17
         Characterization of pressure differential equation 18
         Total velocity for incompressible case 18
      Saturation differential equation 19
         Characterization of saturation equation 20
         Diffusion-convection equation 21
         Nature of saturation equation 21
         One-dimensional case 22
Three-phase flow 23
   Introduction 23
   Darcy's law 23
   Conservation of each phase 23
   Differential equations 24
   Alternative form of differential equations 24
Flow with change of phase 24
   The general compositional model 24
      Component balances 25

Differential equation	25
Auxiliary relations	25
The black-oil model	27
Simplified, two-component, hydrocarbon system	27
Differential equations for black-oil model	29
Limited compositional model	30
Two-component hydrocarbon system with volatility	30
Differential equations	31
Summary	32
Nomenclature	33

Chapter 2. ELEMENTARY FINITE DIFFERENCES 35

Introduction	35
First-difference quotients	35
Second-difference quotients	37
Grid systems	38
Block-centered grid	38
Point-centered grid	39
Comparison of the two grids	40
Truncation error	41
Nomenclature	43

Chapter 3. NUMERICAL SOLUTION OF PARABOLIC PROBLEMS IN ONE DEPENDENT VARIABLE 45

The forward-difference equation	45
Stability by harmonic analysis (the von Neumann criterion)	46
Implicit difference equations	49
The backward-difference equation	49
The tridiagonal algorithm	50
The Crank-Nicolson difference equation	52
Other explicit difference equations	53
A time-centered explicit equation	53
The Dufort-Frankel approximation	53
Multidimensional problems	55
Forward-difference equation	55
Implicit difference equations	56
Alternating-direction methods	57
The Peaceman-Rachford method	57
The Douglas-Rachford method	60
The Brian and Douglas methods	62
Nomenclature	64

Chapter 4. NUMERICAL SOLUTION OF FIRST-ORDER HYPERBOLIC PROBLEMS IN ONE DEPENDENT VARIABLE 65

Introduction	65
Difference equations	65
Distance-weighting	65
Time-weighting	66
General form of difference equation	66
Linearization of difference equation	68

Stability	68
Stability of centered-in-distance equations	69
Stability of backward-in-distance equations	70
Stability of forward-in-distance equations	70
Truncation error analysis—numerical dispersion	71
Local truncation error	71
Numerical dispersion	74
Superposition of numerical and physical dispersion	74
Example calculations	75
Purpose and details of calculations	75
Solutions showing numerical dispersion	76
Solutions without numerical dispersion	78
Unstable solutions	80
Summary	81
Nomenclature	82
Chapter 5. NUMERICAL SOLUTION OF ELLIPTIC PROBLEMS IN ONE DEPENDENT VARIABLE	83
Elliptic difference equations	83
Formulation	83
Matrix notation; the structure of the coefficient matrix	84
Direct solution of band matrix equations by factorization	87
Application of band algorithm to two-dimensional problems	90
Standard ordering	90
Nonstandard orderings	91
Iterative methods for solving elliptic problems	91
Point relaxation methods	92
Introduction	92
Southwell relaxation	93
Gauss-Seidel method (method of successive displacements)	94
Successive overrelaxation (SOR)	94
Method of simultaneous displacements (Jacobi method)	94
Matrix representation of point-iteration methods	95
Jacobi method	95
Successive overrelaxation	96
Convergence rate for Jacobi iteration by harmonic analysis	97
Error expansion for Dirichlet boundary conditions, point-centered grid	97
Error expansion for Neumann boundary conditions, point-centered gird	97
Error expansion for Neumann boundary conditions, block-centered grid	98
Convergence analysis for Neumann boundary conditions	98
Effect of anisotropy on convergence rate for Neumann boundary conditions	100
Convergence rate for Dirichlet boundary conditions	100
Convergence rate for Jacobi iteration by eigenvalue analysis	101
Definition of eigenvalues and eigenvectors	101
Relation between convergence and eigenvalues	101
Application to Jacobi iteration	102
Convergence rate for successive overrelaxation	103
Property A	104
Eigenvalues of SOR iteration matrix	104
More on eigenvalues of Jacobi iteration matrix	107
Relation between convergence of SOR and convergence of Jacobi method	108
Convergence rate of method of successive displacements	108
Optimum parameter for SOR	109

XII

Comparison of convergence rates	110
Effects of anisotropy and boundary conditions on convergence rate of SOR	112
Line relaxation methods	112
Introduction	112
Line simultaneous displacements (line-Jacobi)	112
Line successive overrelaxation (LSOR)	113
Convergence rate of line-Jacobi iteration	113
With Neumann boundary condtions	113
With Dirichlet boundary conditions	115
Acceleration of convergence with Neumann boundary conditions	115
1-D method of additive corrections	115
2-D method of additive corrections	118
Convergence rates of LSOR and LSORC	118
With Neumann boundary conditions	119
With Dirichlet boundary conditions	119
Summary of convergence rates for point and line relaxation methods	119
Alternating-direction iteration (A.D.I.)	120
Formulation of Peaceman-Rachford iteration procedure	120
Convergence analysis	121
Choice of parameters	121
Optimal parameters and convergence rate for ideal case	124
Variable ΔX and ΔY	126
Other alternating-direction iteration procedures	127
Strongly implicit procedure (S.I.P.)	128
Approximate factorization	128
Choice of coefficients	131
Simple method	131
Method of Dupont, Kendall and Rachford	132
Method of Stone (S.I.P.)	132
Summary	134
Nomenclature	135

Chapter 6. NUMERICAL SOLUTION OF TWO-PHASE FLOW PROBLEMS 139

Introduction	139
Differential equations	139
Basic equations in terms of phase pressures	139
Alternative equations	140
Difference notation	141
Difference operators	141
Injection-production terms	142
Interval absolute permeabilities	143
Mobility weighting	143
Some special combinations	143
Simultaneous numerical solution	144
Explicit difference equations	144
Alternating-direction implicit procedure	144
Simultaneous implicit procedure	146
Stability analysis	147
Solution by alternating-direction iteration	148
Solution by strongly implicit procedure	150
Direct solution	151
Calculation of nonlinear coefficient, S'	151

Material balance	152
Summary	153
Sequential numerical solution	153
Introduction	153
Leapfrog method	154
Stability	154
Saturation creep	155
Other sequential solution methods	156
Sequential solution method using total velocity	157
Implicit and semi-implicit mobilities	160
Limitations resulting from use of explicit mobilities	160
Implicit mobilities	161
Semi-implicit mobilities	162
Numerical dispersion	164
Well rates	164
Injection wells	164
Production wells	164
Explicit production rates	165
Semi-implicit production rates	165
Allocation of total injection or production rates	165
Special rate routines	166
Nomenclature	166
REFERENCES	169
SUBJECT INDEX	173

CHAPTER 1

DIFFERENTIAL EQUATIONS FOR FLOW IN RESERVOIRS

INTRODUCTION

By reservoir simulation, we mean the process of inferring the behavior of a real reservoir from the performance of a *model* of that reservoir. The model may be physical, such as a scaled laboratory model, or *mathematical*. For our purposes, a mathematical model of a physical system is a set of partial differential equations, together with an appropriate set of boundary conditions, which we believe adequately describes the significant physical processes taking place in that system. The processes occurring in petroleum reservoirs are basically fluid flow and mass transfer. Up to three immiscible phases (water, oil, and gas) flow simultaneously, while mass transfer may take place between the phases (chiefly between the gas and oil phases). Gravity, capillary, and viscous forces all play a role in the fluid flow process.

The model equations must account for all these forces, and should also take into account an arbitrary reservoir description with respect to heterogeneity and geometry. The differential equations are obtained by combining Darcy's law for each phase with a simple differential material balance for each phase. In this first chapter, we shall first derive the simple differential equation that describes single-phase flow, and then proceed stepwise to derive the set of simultaneous differential equations that describes the most complicated case of multidimensional, multicomponent, three-phase flow.

In the course of deriving these differential equations, we will be introducing many of the basic concepts of reservoir mechanics. For a fuller treatment of the subject of flow of fluids through porous media, the reader is referred to Collins (1961).

To use differential equations for predicting the behavior of a reservoir, it is necessary to solve them subject to the appropriate boundary conditions. Only for the simplest cases involving homogeneous reservoirs and very regular boundaries (such as a circular boundary about a single well) can solutions be obtained by the classical methods of mathematical physics. Numerical methods carried out on high-speed computers, on the other hand, are extremely general in their applicability and have proved to be highly successful for obtaining solutions to very complex reservoir situations. A *numerical model* of a reservoir, then, is a computer program that uses numerical methods to obtain an approximate solution to the mathematical model.

Over the past decade, numerical reservoir simulation has gained wide acceptance throughout the petroleum industry, a consequence of three major factors: a) a tremendous increase in computing speed and capacity, resulting in lower unit computing cost; b) improvements in the numerical algorithms for solving the partial differential equations; and, perhaps most important, c) the generality built into reservoir simulators permitting them to model more realistically the wide variety of oil and gas reservoirs that exist throughout the world.

The remaining chapters of this book will be devoted to outlining some of the basic ideas involved in the numerical solution of partial differential equations by finite-difference methods, and to showing their application to the equations describing reservoir behavior.

SINGLE-PHASE FLOW

Darcy's law

Darcy's law for single-phase flow states that in a horizontal system the volumetric flow rate, $\hat{Q}$, through a sample of porous material of length L and a cross-sectional area A, is given by:

$$\hat{Q} = \frac{KA}{\mu}\frac{\Delta p}{L} \tag{1-1}$$

where Δp is the applied pressure drop across the sample, μ is the viscosity of the fluid, and K is the absolute permeability of the medium. For flow in only one direction (say, parallel to the x-axis), we can write Darcy's law in the following differential form:

$$v = \frac{\hat{Q}}{A} = -\frac{K}{\mu}\frac{\partial p}{\partial x} \tag{1-2}$$

where $v = \hat{Q}/A$ is a superficial flow *velocity*, and $\partial p/\partial x$ is the pressure gradient in the x-direction. Note the negative sign in eq. (1-2), which indicates that the pressure declines in the direction of flow.

The differential form of Darcy's law may be generalized to three dimensions as follows:

$$v_x = -\frac{K}{\mu}\frac{\partial p}{\partial x} \tag{1-3a}$$

$$v_y = -\frac{K}{\mu}\frac{\partial p}{\partial y} \tag{1-3b}$$

$$v_z = -\frac{K}{\mu}\frac{\partial p}{\partial z} \tag{1-3c}$$

where v_x, v_y, and v_z are the x-, y-, and z-components of a velocity vector, $\vec{v}$, oriented in some arbitrary direction in three-dimensional space. Equations (1-3) do not take gravity into account, however, and must be modified to include gravity terms. For generality, we shall take the depth, D, to be an arbitrary function of the coordinates, (x, y, z), rather than commit ourselves at the outset to identifying any one of the coordinates (frequently z) with the depth. Then the differential form of Darcy's law becomes:

$$v_x = -\frac{K}{\mu}\left(\frac{\partial p}{\partial x} - \rho g \frac{\partial D}{\partial x}\right) \tag{1-4a}$$

$$v_y = -\frac{K}{\mu}\left(\frac{\partial p}{\partial y} - \rho g \frac{\partial D}{\partial y}\right) \tag{1-4b}$$

$$v_z = -\frac{K}{\mu}\left(\frac{\partial p}{\partial z} - \rho g \frac{\partial D}{\partial z}\right) \tag{1-4c}$$

where ρ is the density of the fluid and g is the acceleration due to gravity.

One-dimensional, single-phase, compressible flow

In deriving a differential equation for flow in one dimension, we wish to include the possibilities that the cross-sectional area for flow, A, as well as the depth, D, are functions of the single space variable, x. We shall include a term for injection of fluid by using a variable, q, equal to the *mass* rate of injection per unit volume of reservoir. (A negative q implies production.) Finally, we shall recognize that the density of the fluid will be changing with time. (Frequently, the porosity of the medium, ϕ, can also change with time.)

Consider a mass balance about the small box shown in Fig. 1A. The box has length Δx; the left face has area $A(x)$, the right face has area $A(x + \Delta x)$. The rate at which fluid mass enters the box at the left face is given by:

$$\rho(x) \cdot v_x(x) \cdot A(x) = (A\rho v_x)_x$$

The rate at which fluid leaves at the right face is:

$$\rho(x + \Delta x) \cdot v_x(x + \Delta x) \cdot A(x + \Delta x) = (A\rho v_x)_{x+\Delta x}$$

The volume of the box is $\bar{A}\Delta x$. Here $\bar{A}$ indicates the average value of A between x and $x + \Delta x$. Then the rate at which fluid mass is injected into the box is:

$$\bar{q}\bar{A}\Delta x$$

The mass contained in the box is $\bar{\phi}\bar{\rho}\bar{A}\Delta x$. Then the *rate of accumulation* of mass in the box is:

$$\frac{\partial(\bar{\phi}\bar{\rho})}{\partial t}\bar{A}\Delta x$$

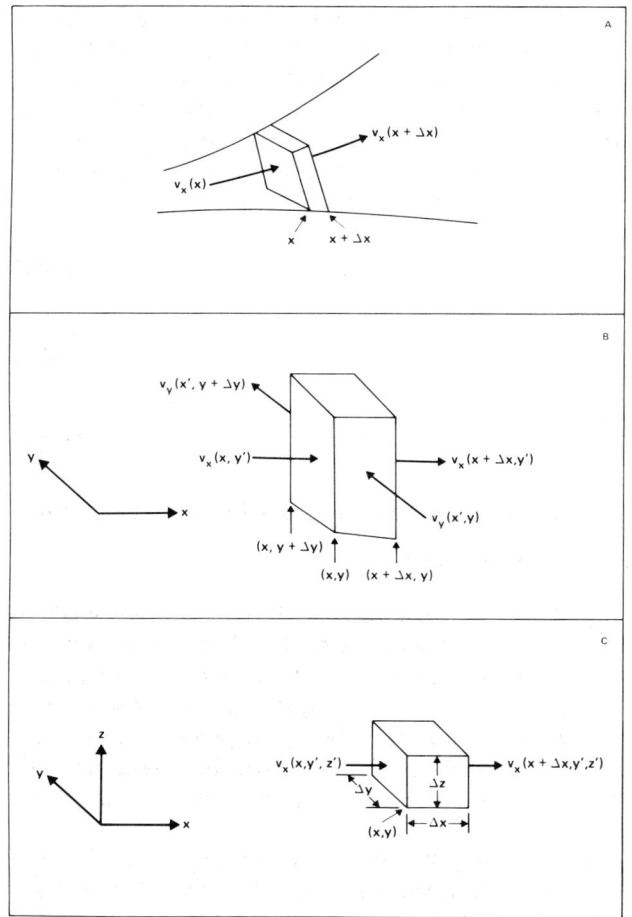

Fig. 1. Differential elements of volume. A. For one-dimensional flow. B. For two-dimensional flow. C. For three-dimensional flow.

Since mass must be *conserved*, we have:

[rate in] − [rate out] + [rate injected] = [rate of accumulation] (1-5)

Thus:

$$(A\rho v_x)_x - (A\rho v_x)_{x+\Delta x} + \bar{q}\bar{A}\Delta x = \bar{A}\frac{\partial(\bar{\phi}\bar{\rho})}{\partial t}\Delta x$$

Dividing by Δx gives:

$$-\frac{(A\rho v_x)_{x+\Delta x} - (A\rho v_x)_x}{\Delta x} + \bar{A}\bar{q} = \bar{A}\frac{\partial(\bar{\phi}\bar{\rho})}{\partial t}$$

Taking the limit as $\Delta x \to 0$ (and noting that a derivative is defined by a limit):

$$\frac{\partial f}{\partial x}(x) = \lim_{\Delta x \to 0} \frac{f(x + \Delta x) - f(x)}{\Delta x}$$

and that $\bar{A} \to A(x)$, $\bar{\rho} \to \rho(x)$, etc., we obtain:

$$-\frac{\partial(A\rho v_x)}{\partial x} + Aq = A\frac{\partial(\phi\rho)}{\partial t} \tag{1-6}$$

Two-dimensional, single-phase, compressible flow

In two-dimensional flow, we wish to allow for the variation of the thickness of the reservoir, $H = H(x, y)$. Consider now a mass balance about the small box in Fig. 1B. The box has length Δx and width Δy. The center of the box is located at $x' = x + \frac{1}{2}\Delta x$ and $y' = y + \frac{1}{2}\Delta y$. The left face has area $H(x, y') \cdot \Delta y$. Hence the rate at which fluid mass enters the box at the left face is given by:

$$\rho(x, y') \cdot v_x(x, y') \cdot H(x, y') \cdot \Delta y = \Delta y (H\rho v_x)_{x, y'}$$

Similarly, fluid mass leaves the box at the right face at the rate:

$$\Delta y (H\rho v_x)_{x+\Delta x, y'}$$

Fluid enters the front face at the rate:

$$\Delta x (H\rho v_y)_{x', y}$$

and leaves the rear face at the rate:

$$\Delta x (H\rho v_y)_{x', y+\Delta y}$$

As the volume of the box is $\bar{H} \cdot \Delta x \cdot \Delta y$, the rate at which fluid mass is injected into the box is:

$$\bar{q}\bar{H}\Delta x \Delta y$$

and the rate of accumulation of mass in the box is:

$$\frac{\partial(\bar{\phi}\bar{\rho})}{\partial t}\bar{H}\Delta x \Delta y$$

Substitution into eq. (1-5) gives:

$$[\Delta y(H\rho v_x)_{x, y'} + \Delta x(H\rho v_y)_{x', y}] - [\Delta y(H\rho v_x)_{x+\Delta x, y'} + \Delta x(H\rho v_y)_{x', y+\Delta y}]$$

$$+ [\bar{q}\bar{H}\Delta x \Delta y] = \bar{H}\frac{\partial(\bar{\phi}\bar{\rho})}{\partial t}\Delta x \Delta y$$

By rearranging, dividing by $\Delta x\ \Delta y$, and taking the limits as $\Delta x \to 0$ and $\Delta y \to 0$, we obtain:

$$-\frac{\partial(H\rho v_x)}{\partial x} - \frac{\partial(H\rho v_y)}{\partial y} + Hq = H\frac{\partial(\phi\rho)}{\partial t} \tag{1-7}$$

Three-dimensional, single-phase, compressible flow

Consider now a mass balance about the small box in Fig. 1C, with length Δx, width Δy, and height Δz. The center of the box is located at $x' = x + \frac{1}{2}\Delta x$, $y' = y + \frac{1}{2}\Delta y$, and $z' = z + \frac{1}{2}\Delta z$. The area of the left face is $\Delta y \Delta z$; hence the rate at which fluid mass enters the box at the left face is:

$$\rho(x, y', z') \cdot v_x(x, y', z') \cdot \Delta y \Delta z = \Delta y \Delta z (\rho v_x)_{x, y', z'}$$

Fluid leaves the right face at the rate:

$$\Delta y \Delta z (\rho v_x)_{x+\Delta x, y', z'}$$

Fluid enters the front face at the rate:

$$\Delta x \Delta z (\rho v_y)_{x', y, z'}$$

and leaves the rear face at the rate:

$$\Delta x \Delta z (\rho v_y)_{x', y+\Delta y, z'}$$

Fluid enters the bottom face at the rate:

$$\Delta x \Delta y (\rho v_z)_{x', y', z}$$

and leaves the top face at the rate:

$$\Delta x \Delta y (\rho v_z)_{x', y', z+\Delta z}$$

The volume of the box is $\Delta x \cdot \Delta y \cdot \Delta z$; thus the rate of injection of mass into the box is:

$$\bar{q} \Delta x \Delta y \Delta z$$

and the rate of accumulation of mass in the box is:

$$\frac{\partial(\bar{\phi}\bar{\rho})}{\partial t} \Delta x \Delta y \Delta z$$

By substituting these rates into eq. (1-5), dividing by $\Delta x \Delta y \Delta z$, and taking the limits as $\Delta x \to 0$, $\Delta y \to 0$, and $\Delta z \to 0$, we obtain:

$$-\frac{\partial(\rho v_x)}{\partial x} - \frac{\partial(\rho v_y)}{\partial y} - \frac{\partial(\rho v_z)}{\partial z} + q = \frac{\partial(\phi\rho)}{\partial t} \tag{1-8}$$

Differential operators

Let u_x, u_y, and u_z be the x-, y-, and z-components of a vector $\vec{u}$. The *divergence* of this vector, written $\nabla \cdot \vec{u}$, is a differential operator on $\vec{u}$ such that:

$$\nabla \cdot \vec{u} = \frac{\partial u_x}{\partial x} + \frac{\partial u_y}{\partial y} + \frac{\partial u_z}{\partial z} \tag{1-9}$$

Another useful differential operator is the gradient, which is a vector formed from the derivatives of a scalar function. If u is a scalar function, the gradient, written ∇u, is a vector whose x-component is $\partial u/\partial x$, whose y-component is $\partial u/\partial y$, and whose z-component is $\partial u/\partial z$. Thus $\partial p/\partial x$, $\partial p/\partial y$, and $\partial p/\partial z$, are the components of ∇p, and $\partial D/\partial x$, $\partial D/\partial y$, and $\partial D/\partial z$ are the components of ∇D. We can now write Darcy's law (eqs. (1-4)) in the compact form:

$$\vec{v} = -\frac{K}{\mu}(\nabla p - \rho g \nabla D) \tag{1-10}$$

A combination of operators which we shall see very frequently is $\nabla \cdot (f \nabla u)$, where f and u are both scalar functions. Since $f\, \partial u/\partial x$, $f\, \partial u/\partial y$, and $f\, \partial u/\partial z$ are the x-, y-, and z-components of the vector $(f \nabla u)$,

$$\nabla \cdot (f \nabla u) = \frac{\partial}{\partial x}\left(f \frac{\partial u}{\partial x}\right) + \frac{\partial}{\partial y}\left(f \frac{\partial u}{\partial y}\right) + \frac{\partial}{\partial z}\left(f \frac{\partial u}{\partial z}\right) \tag{1-11}$$

A special case of this combination is the divergence of the gradient of a function, i.e., $\nabla \cdot (\nabla u)$. This is called the Laplacian of u, and is abbreviated as $\nabla^2 u$. By setting $f \equiv 1$ in eq. (1-11), we obtain:

$$\nabla^2 u = \frac{\partial^2 u}{\partial x^2} + \frac{\partial^2 u}{\partial y^2} + \frac{\partial^2 u}{\partial z^2} \tag{1-12}$$

General equation for single-phase flow

It will be most convenient in our later work if we can use the same differential equation for flow with any number of dimensions. This we can do by arbitrarily defining a "geometric factor" function, α, as follows:

one dimension: $\alpha(x, y, z) = A(x)$ Area   (1-13a)

two dimensions: $\alpha(x, y, z) = H(x, y)$ height   (1-13b)

three dimensions: $\alpha(x, y, z) \equiv 1$ (1-13c)

Then eqs. (1-6), (1-7), and (1-8) can each be written as:

$$-\frac{\partial(\alpha \rho v_x)}{\partial x} - \frac{\partial(\alpha \rho v_y)}{\partial y} - \frac{\partial(\alpha \rho v_z)}{\partial z} + \alpha q = \alpha \frac{\partial(\phi \rho)}{\partial t} \tag{1-14}$$

We recognize that in two-dimensional flow v_z is assumed to be zero, while in one-dimensional flow both v_y and v_z are assumed to be zero. In this way we complete the correspondence between eq. (1-14) and eqs. (1-6) and (1-7).

As α and ρ are scalar functions, then $(\alpha\rho v_x)$, $(\alpha\rho v_y)$, and $(\alpha\rho v_z)$ are the x-, y-, and z-components of the vector $(\alpha\rho\vec{v})$. Using the definition of divergence, eq. (1-14) can be written in the very compact form:

$$-\nabla \cdot (\alpha\rho\vec{v}) + \alpha q = \alpha \frac{\partial(\phi\rho)}{\partial t} \tag{1-15}$$

Finally, by substituting eq. (1-10) into (1-15), one general equation for single-phase flow is obtained:

$$\nabla \cdot \left[\frac{\alpha\rho K}{\mu}(\nabla p - \rho g \nabla D) \right] + \alpha q = \alpha \frac{\partial(\phi\rho)}{\partial t} \tag{1-16}$$

In addition to specifying boundary conditions, it is necessary also to specify porosity and an equation of state for the fluid; that is, relationships between porosity, density and pressure:

$$\phi = \phi(p), \quad \rho = \rho(p)$$

When this is done we have, in theory, all the information necessary to solve the problem.

Boundary conditions

In reservoir simulation, a frequent boundary condition is that the reservoir lies within some closed curve C across which there is no flow, and fluid injection and production take place at wells which can be represented by point sources and sinks. Strictly, we should represent the no-flow boundary conditon by requiring that the normal component of the vector $\vec{v}$ at the curve C be zero. This is relatively difficult to do numerically for an arbitrary curve. However, there usually is little interest in an accurate solution in the neighborhood of the curved boundary; rather, our interest lies mostly within the interior of the reservoir. For this reason, it is adequate to represent the curved boundary in the crude way shown in Fig. 2. The reservoir is embedded in a rectangle (or rectangular parallelopiped), and the functions K and ϕ are set to zero outside the curve C.

As stated above, a well is equivalent to a point source or sink. Numerically, we cannot represent a true point source or sink, wherein q is zero everywhere except at the wells, and infinite at the location of the wells. Again we resort to an approximation. Let Q be the desired mass rate of injection at a well. Let V be the volume of a small box centered at the well. Within the box, we take:

$$q = Q/V$$

and outside the box, we set q to zero. Inasmuch as we will divide the computing region into a grid with spacing Δx (for 1-D), Δx and Δy (for 2-D),

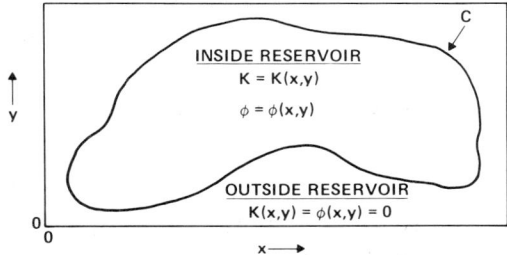

Fig. 2. Representation of an irregularly shaped reservoir.

and Δx, Δy, and Δz (for 3-D), we choose $V = A\Delta x$ (for 1-D), $V = H\Delta x \Delta y$ (for 2-D), or $V = \Delta x \Delta y \Delta z$ (for 3-D).

There are occasions, of course, when we wish part or all of the boundary of the reservoir to be straight. In such a case, a no-flow boundary condition is easy to represent by specifying that the normal component of $\vec{v}$ be zero. Flow across a boundary can be represented by specifying the value of the normal component of $\vec{v}$; an easy, alternative stratagem is to continue to require no flow across the boundary and to place fictitious wells at grid points on or near the boundary. Injection at such a fictitious well represents flow into the region across the boundary, while production would represent flow out of the region.

Finally, it is desired to include among possible boundary conditions the specification of pressure, rather than rate, either at curved or straight boundaries or at wells (i.e., at points within the interior).

Special cases of single-phase flow

It is of interest to examine several simplifications of eq. (1-16) — simplifications that lead to classical equations of mathematical physics. While these simple equations are not directly candidates for numerical solution by reservoir simulators, they will serve as excellent prototypes for explaining the numerical methods used in the simulators.

The initial simplification is to assume that gravitational effects are negligible and that the flow region is free of sources and sinks. With $\nabla D = 0$, and $q = 0$, eq. (1-16) can be written:

$$\nabla \cdot \left(\frac{\alpha \rho K}{\mu} \nabla p \right) = \alpha \frac{\partial (\phi \rho)}{\partial t} \qquad (1\text{-}17)$$

Several different cases arise, depending on the choice of the equation of state for the fluid.

Ideal liquid of constant compressibility

The compressibility of a fluid, c, is defined, for isothermal conditions, by:

$$c = \frac{1}{\rho}\frac{d\rho}{dp} \tag{1-18}$$

For an *ideal liquid*, that is, one with constant compressibility (as well as constant viscosity), integration yields:

$$\rho = \rho_o \exp[c(p - p_o)] \tag{1-19}$$

where ρ_o is the density at the reference pressure, p_o. This particular equation of state applies rather well to most liquids, unless they contain large quantities of dissolved gas. From eq. (1-18):

$$\rho dp = \frac{d\rho}{c}$$

$$\rho \nabla p = \frac{1}{c}\nabla \rho$$

and eq. (1-17) becomes:

$$\nabla \cdot \left(\frac{\alpha K}{\mu c}\nabla \rho\right) = \alpha \frac{\partial(\phi\rho)}{\partial t}$$

If, in addition, the porous medium and the reservoir are homogeneous, then α, K, and ϕ are uniform, and we obtain:

$$\nabla^2 \rho = \frac{\phi \mu c}{K}\frac{\partial \rho}{\partial t} \tag{1-20}$$

This is exactly of the same form as the Fourier equation of heat conduction discussed below.

Liquids of slight compressibility

For liquids of *slight* compressibility, the same equation as (1-20) is obtained in terms of pressure:

$$\nabla^2 p = \frac{\phi \mu c}{K}\frac{\partial p}{\partial t} \tag{1-21}$$

When analytical techniques, such as Laplace transforms, are used for the solution of the heat conduction equation to simulate reservoir performance, eq. (1-21) is most frequently used, rather than (1-20). In doing so, it is important to recognize the significance of the assumption that compressibility is small. To demonstrate the nature of this assumption, we rewrite eq. (1-18) as $d\rho = c\rho dp$ and obtain:

$$\frac{\partial \rho}{\partial t} = c\rho \frac{\partial p}{\partial t} \tag{1-22}$$

and:

$$\frac{\partial \rho}{\partial x} = c\rho \frac{\partial p}{\partial x} \tag{1-23}$$

Differentiating eq. (1-23) with respect to x gives:

$$\frac{\partial^2 \rho}{\partial x^2} = c\rho \frac{\partial^2 p}{\partial x^2} + c\frac{\partial \rho}{\partial x}\frac{\partial p}{\partial x} = c\rho \frac{\partial^2 p}{\partial x^2} + c^2\rho \left(\frac{\partial p}{\partial x}\right)^2$$

Similar expressions hold for $\partial^2 \rho/\partial y^2$ and $\partial^2 \rho/\partial z^2$. Thus:

$$\nabla^2 \rho = c\rho \nabla^2 p + c^2 \rho \left[\left(\frac{\partial p}{\partial x}\right)^2 + \left(\frac{\partial p}{\partial y}\right)^2 + \left(\frac{\partial p}{\partial z}\right)^2\right] \tag{1-24}$$

Substituting eqs. (1-22) and (1-24) into (1-20), and dividing by $c\rho$, gives:

$$\nabla^2 p + c\left[\left(\frac{\partial p}{\partial x}\right)^2 + \left(\frac{\partial p}{\partial y}\right)^2 + \left(\frac{\partial p}{\partial z}\right)^2\right] = \frac{\phi \mu c}{K}\frac{\partial p}{\partial t} \tag{1-25}$$

For eq. (1-21) to be a good approximation for (1-25) (and thus for (1-20)), it is necessary that:

$$\frac{\partial^2 p}{\partial x^2} + \frac{\partial^2 p}{\partial y^2} + \frac{\partial^2 p}{\partial z^2} \gg c\left[\left(\frac{\partial p}{\partial x}\right)^2 + \left(\frac{\partial p}{\partial y}\right)^2 + \left(\frac{\partial p}{\partial z}\right)^2\right]$$

Hence the requirement that c be very small.

Ideal gas

In the case of flow of an *ideal gas* through porous media, we can again ignore gravitational effects. The equation of state is:

$$\rho = \frac{M}{RT}p \tag{1-26}$$

where M is molecular weight, R the gas constant, and T is the absolute temperature. Substitution of eq. (1-26) into (1-17) gives:

$$\nabla \cdot \left(\frac{\alpha K}{\mu} p \nabla p\right) = \alpha \frac{\partial(\phi p)}{\partial t}$$

If α, K, ϕ, and μ are again considered constant, this can be written in the very simple form:

$$\nabla^2 p^2 = \frac{\phi \mu}{Kp}\frac{\partial p^2}{\partial t} \tag{1-27}$$

While eq. (1-27) is nonlinear, it is quite similar to (1-21).

Incompressible flow

An important category of single-phase flow problems is that of incompressible flow. With ρ and ϕ constant, eq. (1-16) becomes:

$$\nabla \cdot \left[\frac{\alpha K}{\mu} (\nabla p - \rho g \nabla D) \right] + \frac{\alpha q}{\rho} = 0 \tag{1-28}$$

It is convenient to define a potential, Φ, as:

$$\Phi = p - \rho g D$$

Then eq. (1-28) becomes:

$$\nabla \cdot \left(\frac{\alpha K}{\mu} \nabla \Phi \right) + \frac{\alpha q}{\rho} = 0$$

For the case of a homogeneous reservoir (constant α and K) as well as constant viscosity, this simplifies to Poisson's equation (see below):

$$\nabla^2 \Phi + \frac{q\mu}{K\rho} = 0$$

and, in the region where there are no fluid sources or sinks, to Laplace's equation (also see below):

$$\nabla^2 \Phi = 0$$

For sufficiently regular boundary conditions, the results of potential theory (which involve solution of Laplace's equation) can then be applied to some simple reservoir simulation problems.

TYPES OF SECOND-ORDER DIFFERENTIAL EQUATIONS

There are basically three types of second-order differential equations, parabolic, elliptic, and hyperbolic. We need to be able to distinguish among these types, since the numerical methods for their solution need to be considered separately, as we shall do in Chapters 3, 4, and 5.

In the discussion below, the variable, u, is a generalized dependent variable.

Parabolic equations

The prototype parabolic equation is the Fourier equation of diffusion or heat conduction,

$$\nabla^2 u = \frac{1}{k} \frac{\partial u}{\partial t} \tag{1-29}$$

which arises in the theory of diffusion, conduction of heat, and in electrical conduction (in purely resistive material). Here, k is the diffusivity or the thermal (or electrical) conductivity.

Elliptic equations

Laplace's equation, $\nabla^2 u = 0$, and Poisson's equation, $\nabla^2 u = f(x, y, z)$, are the simplest of the elliptic equations. They arise in electrostatic and magnetic field theory and in hydrodynamics of incompressible fluids.

Hyperbolic equations

The prototype hyperbolic equation is the wave equation,

$$\nabla^2 u = \frac{1}{v^2} \frac{\partial^2 u}{\partial t^2} \tag{1-30}$$

which arises in acoustics and electrodynamics. Here, v is the velocity of propagation of the acoustic or electromagnetic disturbance.

Classification of equations

If we restrict ourselves to two independent variables (either x and y, or x and t), then these equations may be written in the following general form:

$$A \frac{\partial^2 u}{\partial x^2} + B \frac{\partial^2 u}{\partial t^2} = f\left(\frac{\partial u}{\partial x}, \frac{\partial u}{\partial t}, u\right) \tag{1-31}$$

This equation may then be classified into the three types as follows:
1) $A \cdot B > 0$. Elliptic
2) $A \cdot B = 0$. Parabolic
3) $A \cdot B < 0$. Hyperbolic

There is an obvious analogy with the familiar algebraic equation:

$$Ax^2 + By^2 = y$$

which is an equation for an ellipse when $A \cdot B > 0$, for a parabola when $B = 0$, and for a hyperbola when $A \cdot B < 0$.

First-order hyperbolic equations

In one space dimension, the second-order hyperbolic eq. (1-30) is:

$$\frac{\partial^2 u}{\partial x^2} - \frac{1}{v^2} \frac{\partial^2 u}{\partial t^2} = 0$$

which can be "factored" into two first-order parts as follows:

$$\left(\frac{\partial}{\partial x} - \frac{1}{v} \frac{\partial}{\partial t}\right)\left(\frac{\partial}{\partial x} + \frac{1}{v} \frac{\partial}{\partial t}\right) u = 0$$

The second factor is equivalent to the first-order *convection* equation:

$$-v\frac{\partial u}{\partial x} = \frac{\partial u}{\partial t} \tag{1-32}$$

Below, we shall see that eq. (1-32) is important in the theory of multiphase flow through porous media. On the other hand, second-order hyperbolic problems will be of little interest to us in this book. Consequently, the discussion of numerical methods for the solution of hyperbolic problems (in Chapter 4) will be focussed entirely on first-order problems.

TWO-PHASE FLOW

Introduction

In reservoir simulation, we are primarily concerned with modelling the displacement, within a porous medium, of oil by either water or gas. While the displacing fluid may be immiscible with the fluid being displaced, the displacement does not take place as a piston-like process with a sharp interface between the two fluids. Rather, simultaneous flow of the two immiscible fluids takes place within the porous medium.

In considering this simultaneous flow we assume, for the present, no mass transfer between the two fluids. One of the fluids wets the porous medium more than the other; we refer to this as the wetting phase fluid (and use the subscript w), and we refer to the other as the nonwetting phase fluid (and use the subscript n). In a water-oil system, water is most often the wetting phase; in an oil-gas system, oil is the wetting phase.

Several new concepts peculiar to multiphase flow must now be introduced, namely saturation, capillary pressure, and relative permeability. The saturation of a phase is defined as the fraction of the void volume of the porous medium filled by that phase. Since the two fluids jointly fill the void space, we have:

$$S_n + S_w = 1 \tag{1-33}$$

Because of surface tension and the curvature of the interfaces between the two fluids within the small pores, the pressure in the nonwetting fluid is higher than the pressure in the wetting fluid. The difference between these two pressures is the capillary pressure, p_c:

$$p_c = p_n - p_w \tag{1-34}$$

We accept, as an empirical fact, that the capillary pressure is a unique function of saturation:

$$p_n - p_w = p_c(S_w) \tag{1-35}$$

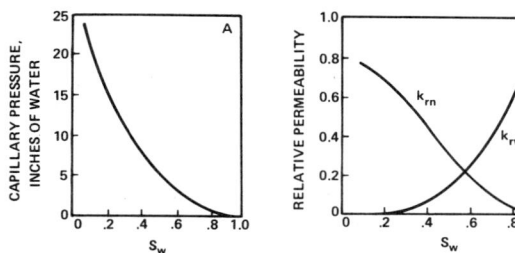

Fig. 3. Typical curves for two-phase data. A. Capillary pressure. B. Relative permeabilities.

A typical capillary pressure curve is shown in Fig. 3A.

Darcy's law

Darcy's law is extended to multiphase flow by postulating that these same phase pressures are involved in causing each fluid to flow. Equation (1-10) can then be written for each fluid:

$$\vec{v}_n = -\frac{K_n}{\mu_n}(\nabla p_n - \rho_n g \nabla D) \qquad (1\text{-}36a)$$

$$\vec{v}_w = -\frac{K_w}{\mu_w}(\nabla p_w - \rho_w g \nabla D) \qquad (1\text{-}36b)$$

Here $\vec{v}_n$ and $\vec{v}_w$ are the superficial velocities for the nonwetting and wetting fluids, respectively, μ_n and μ_w are the respective viscosities, and ρ_n and ρ_w the respective densities. K_n and K_w are the effective permeabilities for flow for each of the two fluids. Because the simultaneous flow of the two fluids causes each to interfere with the flow of the other, these effective permeabilities must be less than or equal to the single-fluid permeability, K, of the medium. Relative permeabilities are therefore defined by:

$$k_{rn} = \frac{K_n}{K} \leqslant 1 \qquad (1\text{-}37a)$$

$$k_{rw} = \frac{K_w}{K} \leqslant 1 \qquad (1\text{-}37b)$$

Again we accept as an empirical fact that these relative permeabilities are unique functions of the saturation. Typical relative permeability curves are shown in Fig. 3B.

We rewrite Darcy's law now, using the relative permeabilities:

$$\vec{v}_n = -\frac{K k_{rn}}{\mu_n}(\nabla p_n - \rho_n g \nabla D) \qquad (1\text{-}38a)$$

$$\vec{v}_w = -\frac{Kk_{rw}}{\mu_w}(\nabla p_w - \rho_w g \nabla D) \tag{1-38b}$$

Conservation of each phase

Except for the accumulation term, the same considerations that led to the derivations of eqs. (1-6), (1-7), and (1-8) apply also in deriving a differential equation of flow for each phase. To obtain the accumulation term, we note that the amount of mass of each phase in a differential volume is the product of the volume of the differential element, the porosity, the density of the phase, *and the saturation of the phase*. Thus the rates of accumulation are:

	Nonwetting phase	Wetting phase
1-D	$A\dfrac{\partial(\phi\rho_n S_n)}{\partial t}\Delta x$	$A\dfrac{\partial(\phi\rho_w S_w)}{\partial t}\Delta x$
2-D	$H\dfrac{\partial(\phi\rho_n S_n)}{\partial t}\Delta x \Delta y$	$H\dfrac{\partial(\phi\rho_w S_w)}{\partial t}\Delta x \Delta y$
3-D	$\dfrac{\partial(\phi\rho_n S_n)}{\partial t}\Delta x \Delta y \Delta z$	$\dfrac{\partial(\phi\rho_w S_w)}{\partial t}\Delta x \Delta y \Delta z$

Hence we extend eq. (1-15), which is the continuity equation for single-phase flow, as follows:

$$-\nabla \cdot (\alpha \rho_n \vec{v}_n) + \alpha q_n = \alpha \frac{\partial(\phi \rho_n S_n)}{\partial t} \tag{1-39a}$$

$$-\nabla \cdot (\alpha \rho_w \vec{v}_w) + \alpha q_w = \alpha \frac{\partial(\phi \rho_w S_w)}{\partial t} \tag{1-39b}$$

The differential equations for two-phase flow

By combining eqs. (1-39) with (1-38), we obtain the set of simultaneous differential equations that describe two-phase flow:

$$\nabla \cdot \left[\frac{\alpha \rho_n K k_{rn}}{\mu_n}(\nabla p_n - \rho_n g \nabla D)\right] + \alpha q_n = \alpha \frac{\partial(\phi \rho_n S_n)}{\partial t} \tag{1-40a}$$

$$\nabla \cdot \left[\frac{\alpha \rho_w K k_{rw}}{\mu_w}(\nabla p_w - \rho_w g \nabla D)\right] + \alpha q_w = \alpha \frac{\partial(\phi \rho_w S_w)}{\partial t} \tag{1-40b}$$

These equations are extremely general in their applicability, including, as they do, the effects of compressibility, capillary pressure, and relative permeability, as well as variations with position of the absolute permeability and the porosity.

Alternative differential equations for two-phase flow

Because of the superficial resemblance of eqs. (1-40) to the heat conduction equation (1-29), one might expect two-phase flow problems to be essentially parabolic in nature. That this is not necessarily so may be demonstrated by examining in some detail an alternative pair of differential equations equivalent to (1-40). The first of this pair is a "pressure equation" that primarily describes how pressure varies with time and position; the second is a "saturation equation" that describes the variation of saturation with time and position.

Pressure differential equation

The primary objective in deriving the pressure differential equation is to eliminate the time derivatives of saturation. To do this, we begin by expanding the time derivatives of eqs. (1-39) to obtain:

$$-\nabla \cdot (\alpha \rho_n \vec{v}_n) + \alpha q_n = \alpha \left[\rho_n S_n \frac{\partial \phi}{\partial t} + \phi S_n \frac{d\rho_n}{dp_n} \frac{\partial p_n}{\partial t} + \phi \rho_n \frac{\partial S_n}{\partial t} \right] \quad (1\text{-}41)$$

$$-\nabla \cdot (\alpha \rho_w \vec{v}_w) + \alpha q_w = \alpha \left[\rho_w S_w \frac{\partial \phi}{\partial t} + \phi S_w \frac{d\rho_w}{dp_w} \frac{\partial p_w}{\partial t} + \phi \rho_w \frac{\partial S_w}{\partial t} \right] \quad (1\text{-}42)$$

We divide eq. (1-41) by $\alpha \rho_n$ and eq. (1-42) by $\alpha \rho_w$, and add the resulting equations. Then, using eq. (1-33), we obtain:

$$-(1/\alpha\rho_n)\nabla \cdot (\alpha\rho_n\vec{v}_n) - (1/\alpha\rho_w)\nabla \cdot (\alpha\rho_w\vec{v}_w) + Q_t$$

$$= \frac{\partial \phi}{\partial t} + \phi S_n c_n \frac{\partial p_n}{\partial t} + \phi S_w c_w \frac{\partial p_w}{\partial t} \quad (1\text{-}43)$$

where:

$$Q_t = (q_n/\rho_n) + (q_w/\rho_w) \quad (1\text{-}44)$$

is the total *volumetric* injection rate, and:

$$c_n = \frac{1}{\rho_n} \frac{d\rho_n}{dp_n} \quad (1\text{-}45a)$$

$$c_w = \frac{1}{\rho_w} \frac{d\rho_w}{dp_w} \quad (1\text{-}45b)$$

are phase compressibilities, analogous to the single-phase compressibility defined by eq. (1-18). Note that time derivatives of saturation are absent from eq. (1-43).

An *average* pressure may be defined by:

$$p_{avg} = (p_n + p_w)/2 \quad (1\text{-}46)$$

The individual phase pressures can then be expressed in terms of the average pressure and the capillary pressure by:

$$p_n = p_{avg} + \tfrac{1}{2} p_c \tag{1-47a}$$

$$p_w = p_{avg} - \tfrac{1}{2} p_c \tag{1-47b}$$

In addition, let us define phase mobilities, λ_n and λ_w, by:

$$\lambda_n = K k_{rn}/\mu_n \tag{1-48a}$$

$$\lambda_w = K k_{rw}/\mu_w \tag{1-48b}$$

Then, substitution of eqs. (1-38) into (1-43) and rearrangement gives the final form of the pressure equation:

$$[(1/\alpha\rho_n)\nabla \cdot (\alpha\rho_n\lambda_n) + (1/\alpha\rho_w)\nabla \cdot (\alpha\rho_w\lambda_w)]\nabla p_{avg} + [(1/2\alpha\rho_n)\nabla \cdot (\alpha\rho_n\lambda_n)$$
$$- (1/2\alpha\rho_w)\nabla \cdot (\alpha\rho_w\lambda_w)]\nabla p_c + Q_t = [(d\phi/dp_{avg})$$
$$+ \phi(S_n c_n + S_w c_w)]\frac{\partial p_{avg}}{\partial t} + [\phi(S_n c_n - S_w c_w)/2]\frac{\partial p_c}{\partial t}$$
$$+ g[(1/\alpha\rho_n)\nabla \cdot (\alpha\rho_n^2\lambda_n) + (1/\alpha\rho_w)\nabla \cdot (\alpha\rho_w^2\lambda_w)]\nabla D \tag{1-49}$$

Characterization of pressure differential equation. In exploring the character of eq. (1-49), we note first that p_c is usually quite small relative to p_{avg}. The final term, involving depth, may be regarded as a modification to the source term, Q_t. Finally, we can ignore for the moment the variation of $\alpha\rho_n$ and $\alpha\rho_w$ with position. Then eq. (1-49) can be simplified to:

$$\nabla \cdot (\lambda_n + \lambda_w)\nabla p_{avg} + Q_t \approx \phi c_t \frac{\partial p_{avg}}{\partial t} \tag{1-50}$$

where c_t is a total compressibility defined by:

$$c_t = (1/\phi)(d\phi/dp_{avg}) + (S_n c_n + S_w c_w) \tag{1-51}$$

Thus we see that eq. (1-50), and therefore eq. (1-49), is basically a parabolic equation. However, while the effects of compressibility may not be ignored in reservoir calculations, they usually do not dominate. Indeed, any reservoir simulator must be capable of dealing satisfactorily with multiphase flow of incompressible fluids, for which case, $c_t = 0$. Thus, as a practical matter, eq. (1-49) must be regarded as being elliptic, or nearly elliptic, in nature.

Total velocity for incompressible case. For the incompressible case, in which ϕ, ρ_n, and ρ_w are constant, eq. (1-43) simplifies to:

$$-(1/\alpha)\nabla \cdot (\alpha \vec{v}_n + \alpha \vec{v}_w) + Q_t = 0$$

If we define a *total* velocity by:

$$\vec{v}_t = \vec{v}_n + \vec{v}_w \tag{1-52}$$

then:

$$\nabla \cdot (\alpha \vec{v}_t) = \alpha Q_t \tag{1-53}$$

The simplicity of this equation indicates the fundamental role that total velocity plays in two-phase flow.

Saturation differential equation

In deriving the saturation equation we can, of course, focus on either the wetting or the nonwetting phase. Here we choose, quite arbitrarily, the wetting phase. Assuming that the solution to eq. (1-49) is known, p_w may be obtained from (1-47b) and then $\vec{v}_w$ obtained from (1-38b). Equation (1-39b), which involves $\vec{v}_w$, could then be used for the saturation equation.

However, a more significant saturation equation can be obtained, one that involves the total velocity defined by eq. (1-52). To this end, we first obtain the wetting phase velocity in terms of the total velocity. From eqs. (1-34), (1-38), and (1-48), we get:

$$\nabla p_c = \nabla p_n - \nabla p_w$$
$$\vec{v}_n = -\lambda_n (\nabla p_n - \rho_n g \nabla D)$$
$$\vec{v}_w = -\lambda_w (\nabla p_w - \rho_w g \nabla D)$$

Combination of these three equations and rearrangement yields:

$$\lambda_n \lambda_w \nabla p_c = -\lambda_w \vec{v}_n + \lambda_n \vec{v}_w + \lambda_n \lambda_w (\rho_n - \rho_w) g \nabla D$$

Using eq. (1-52) to eliminate $\vec{v}_n$, we obtain:

$$(\lambda_n + \lambda_w)\vec{v}_w = \lambda_w \vec{v}_t + \lambda_n \lambda_w [\nabla p_c + (\rho_w - \rho_n)g \nabla D] \tag{1-54}$$

Let us define the following functions of saturation:

$$f_w = \frac{\lambda_w}{\lambda_n + \lambda_w} \tag{1-55}$$

$$h_w = -\frac{\lambda_n \lambda_w}{\lambda_n + \lambda_w} \frac{dp_c}{dS_w} \tag{1-56}$$

A typical curve of f_w vs. S_w is shown in Fig. 4. The negative sign is included in the definition of h_w to keep it positive, since p_c is a decreasing function of S_w. Equation (1-54) then becomes:

$$\vec{v}_w = f_w \vec{v}_t - h_w \nabla S_w + \lambda_n f_w (\rho_w - \rho_n) g \nabla D \tag{1-57}$$

and eq. (1-39b) can be written in the following final form for the saturation equation:

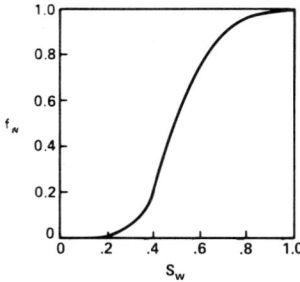

Fig. 4. Typical curve of f_w vs. S_w.

$$\nabla \cdot (\alpha \rho_w h_w \nabla S_w) - \nabla \cdot (\alpha \rho_w f_w)[\vec{v}_t + \lambda_n(\rho_w - \rho_n)g\nabla D] + \alpha q_w = \alpha \frac{\partial(\phi \rho_w S_w)}{\partial t} \quad (1\text{-}58)$$

Characterization of saturation equation. While eq. (1-58) is relatively complex, the first term (which involves the capillary pressure) strongly suggests that it is essentially parabolic in nature, unless capillary effects are unimportant. In that case, the two center terms that involve velocity and gravity become more important, but their significance is not so obvious. To investigate this, we need to simplify eq. (1-58) by assuming incompressibility. With the assumption that ϕ, ρ_n, and ρ_w are constant, eq. (1-58) becomes:

$$\nabla \cdot (\alpha h_w \nabla S_w) - \nabla \cdot (\alpha f_w \vec{v}_t) - \nabla \cdot (\alpha G_w \nabla D) + \alpha(q_w/\rho_w) = \alpha \phi \frac{\partial S_w}{\partial t} \quad (1\text{-}59)$$

where:

$$G_w = f_w \lambda_n (\rho_w - \rho_n) g \quad (1\text{-}60)$$

is another function of saturation.

The nature of the second term of eq. (1-59) is explored by expanding it as follows:

$$\nabla \cdot (\alpha f_w \vec{v}_t) = f_w \nabla \cdot (\alpha \vec{v}_t) + \alpha \vec{v}_t \cdot \nabla f_w$$

The last term is a dot product, which is defined for any two vectors, $\vec{u}$ and $\vec{v}$, by:

$$\vec{u} \cdot \vec{v} = u_x v_x + u_y v_y + u_z v_z \quad (1\text{-}61)$$

In turn, we can write:

$$\nabla f_w = \frac{df_w}{dS_w} \nabla S_w$$

Now, we are primarily concerned with characterizing the saturation equation in the regions of the reservoir between the sources and sinks. Thus we take $q_w = Q_t = 0$. Equation (1-53) becomes:

$$\nabla \cdot (\alpha \vec{v}_t) = 0 \qquad (1\text{-}62)$$

and eq. (1-59) simplifies to:

$$(1/\alpha)\nabla \cdot (\alpha h_w \nabla S_w) - \frac{df_w}{dS_w}\vec{v}_t \cdot \nabla S_w = \phi\frac{\partial S_w}{\partial t} + (1/\alpha)\nabla \cdot (\alpha G_w \nabla D) \qquad (1\text{-}63)$$

Diffusion-convection equation. Equation (1-63) can be regarded as a non-linear variation of the diffusion-convection equation:

$$\mathcal{D}\nabla^2 C - \vec{v} \cdot \nabla C = \phi\frac{\partial C}{\partial t} \qquad (1\text{-}64)$$

which governs multidimensional miscible displacement. Here $\mathcal{D}$ is diffusivity and C is concentration. While the numerical solution of miscible displacement problems is outside the scope of this book, we shall make some references to this equation in subsequent chapters.

The first term of eq. (1-64) is the diffusion term and, when it dominates, (1-64) behaves like the parabolic heat conduction equation (1-29). On the other hand, when the diffusion term is small, the center term, i.e., the *convection* term, dominates and (1-64) approaches the first-order hyperbolic equation:

$$-\vec{v} \cdot \nabla C = \phi\frac{\partial C}{\partial t} \qquad (1\text{-}65)$$

The first-order character of eq. (1-65) may be made more clear by expanding the left-hand side:

$$-v_x\frac{\partial C}{\partial x} - v_y\frac{\partial C}{\partial y} - v_z\frac{\partial C}{\partial z} = \phi\frac{\partial C}{\partial t} \qquad (1\text{-}66)$$

Equation (1-65) can thus be seen to be a multidimensional equivalent of the one-dimensional convection equation (1-32).

Nature of saturation equation. Referring back to eq. (1-63), we can now see that it may be either parabolic or hyperbolic in nature, depending on the importance of the capillary pressure term relative to the convection term. When capillary pressure effects dominate, h_w is large, and (1-63) behaves like a parabolic problem. When capillary pressure effects are small or absent or, more importantly sometimes, when velocities are large, then the convection term dominates, and (1-63) approaches the first-order nonlinear hyperbolic equation:

$$-\frac{df_w}{dS_w}\vec{v}_t \cdot \nabla S_w = \phi\frac{\partial S_w}{\partial t} + (1/\alpha)\nabla \cdot (\alpha G_w \nabla D) \qquad (1\text{-}67)$$

One-dimensional case. Of particular interest is the case of linear displacement in a thin uniform tube inclined upward in the positive x-direction at angle α_d to the horizontal. Then:

$$\nabla D = \frac{\partial D}{\partial x} = -\sin \alpha_d$$

and eq. (1-67) reduces to:

$$-\frac{df_w}{dS_w} v_{tx} \frac{\partial S_w}{\partial x} = \phi \frac{\partial S_w}{\partial t} - \sin \alpha_d \frac{dG_w}{dS_w} \frac{\partial S_w}{\partial x} \tag{1-68}$$

Equation (1-62) reduces to:

$$A(\partial v_{tx}/\partial x) = 0$$

which implies that v_{tx} is constant. Upon substitution of eq. (1-60), eq. (1-68) can be rearranged to:

$$v_{tx} \frac{dF_w}{dS_w} \frac{\partial S_w}{\partial x} = -\phi \frac{\partial S_w}{\partial t} \tag{1-69}$$

where:

$$F_w = f_w[1 - (\lambda_n/v_{tx})(\rho_w - \rho_n)g \sin \alpha_d] \tag{1-70}$$

is the fraction of the flowing stream which is the wetting phase (Collins, 1961, pp. 143—144).

The rate of advance of a point of constant saturation may be derived by using the relation:

$$\left(\frac{dx}{dt}\right)_{S_w} = -\frac{\partial S_w}{\partial t}\left(\frac{\partial S_w}{\partial x}\right)^{-1}$$

Using eq. (1-69) to eliminate $\partial S_w/\partial t$ from this equation, we obtain:

$$\left(\frac{dx}{dt}\right)_{S_w} = \frac{v_{tx}}{\phi} \frac{dF_w}{dS_w} \tag{1-71}$$

This equation, derived by Buckley and Leverett (1942), can be used to solve directly for saturation as a function of position and time. This technique is referred to as the method of characteristics, and is of general utility for solving hyperbolic problems.

Lately, the "Buckley-Leverett equation" has taken on the more general meaning of a saturation equation that involves the total velocity. It may also include the capillary pressure. Thus eqs. (1-63) and (1-67) are referred to as multidimensional forms of the Buckley-Leverett equation.

THREE-PHASE FLOW

Introduction

We now consider the simultaneous flow of three immiscible fluids through porous media. Specifically, the fluids will be gas, oil, and water, and we shall use the subscripts g, o, and w to refer to the gas, oil, and water phases, respectively. Again assume, for the present, no mass transfer between the three fluids. (This frequently unrealistic assumption will be removed later.)

The development of the differential equations for three phases parallels that for two phases. First we have:

$$S_g + S_o + S_w = 1 \tag{1-72}$$

Two independent capillary pressures can be defined:

$$P_{c_{ow}} = p_o - p_w \tag{1-73a}$$

$$P_{c_{go}} = p_g - p_o \tag{1-73b}$$

It is not necessary to define a third capillary pressure, as this would be a simple combination of the other two. That is,

$$P_{c_{gw}} = p_g - p_w = P_{c_{go}} + P_{c_{ow}} \tag{1-73c}$$

There are little experimental data on three-phase capillary pressures, making it necessary to obtain estimates from two-phase data measured on the water-oil and gas-oil subsystems. Peery and Herron (1969) present the schemes used in their program to estimate three-phase capillary pressure and relative permeabilities.

Darcy's law

This is written in the usual way for each of the three phases:

$$\vec{v}_g = -\frac{Kk_{rg}}{\mu_g}(\nabla p_g - \rho_g g \nabla D) \tag{1-74a}$$

$$\vec{v}_o = -\frac{Kk_{ro}}{\mu_o}(\nabla p_o - \rho_o g \nabla D) \tag{1-74b}$$

$$\vec{v}_w = -\frac{Kk_{rw}}{\mu_w}(\nabla p_w - \rho_w g \nabla D) \tag{1-74c}$$

Conservation of each phase

$$-\nabla \cdot (\alpha \rho_g \vec{v}_g) + \alpha q_g = \alpha \frac{\partial(\phi \rho_g S_g)}{\partial t} \tag{1-75a}$$

$$-\nabla \cdot (\alpha \rho_o \vec{v}_o) + \alpha q_o = \alpha \frac{\partial(\phi \rho_o S_o)}{\partial t} \tag{1-75b}$$

$$-\nabla \cdot (\alpha \rho_w \vec{v}_w) + \alpha q_w = \alpha \frac{\partial(\phi \rho_w S_w)}{\partial t} \tag{1-75c}$$

Differential equations

$$\nabla \cdot \left[\frac{\alpha \rho_g K k_{rg}}{\mu_g} (\nabla p_g - \rho_g g \nabla D) \right] + \alpha q_g = \alpha \frac{\partial(\phi \rho_g S_g)}{\partial t} \tag{1-76a}$$

$$\nabla \cdot \left[\frac{\alpha \rho_o K k_{ro}}{\mu_o} (\nabla p_o - \rho_o g \nabla D) \right] + \alpha q_o = \alpha \frac{\partial(\phi \rho_o S_o)}{\partial t} \tag{1-76b}$$

$$\nabla \cdot \left[\frac{\alpha \rho_w K k_{rw}}{\mu_w} (\nabla p_w - \rho_w g \nabla D) \right] + \alpha q_w = \alpha \frac{\partial(\phi \rho_w S_w)}{\partial t} \tag{1-76c}$$

Alternative form of differential equations

As for the two-phase case, it is possible to derive an alternative system of equations equivalent to the three equations of (1-76). This would parallel the development of eqs. (1-49) and (1-58) from eqs. (1-40).

For the three-phase case, the time derivatives of saturations can be eliminated to obtain a single pressure equation. Again, a total velocity can be defined, which is the sum of the three individual phase velocities. Finally, however, instead of a single saturation equation, two simultaneous saturation equations are obtained.

While this alternative system of equations will not be presented here, we can state conclusions regarding their character which are similar to those obtained for the two-phase case. The pressure equation is elliptic or near-elliptic. The saturation equations are parabolic if the capillary pressure effects dominate; otherwise they are hyperbolic or near-hyperbolic.

FLOW WITH CHANGE OF PHASE

The general compositional model

In this section we consider the very general case where there are N chemical species, or components, each of which may exist in any or all of the three phases (gas, oil, and water). Let C_{ig} be the mass fraction of the ith component in the gas phase, C_{io} the mass fraction of the ith component in the oil phase, and C_{iw} the mass fraction of the ith component in the water phase. With this generality, we include not only the distribution of hydrocarbon components between the oil and gas phases, but also allow for a component (such as carbon dioxide) that can also dissolve in the water phase. It would also allow for vaporization of water into the gas phase, which plays an important role in a steam drive process.

Component balances

We can no longer say that the mass of each phase is conserved, as we have done in previous sections, because of the possibility of transfer of various components between the phases. Instead, we observe that the total mass of each component must be conserved. As the mass flux densities for each of the phases are $\rho_g \vec{v}_g$, $\rho_o \vec{v}_o$, and $\rho_w \vec{v}_w$ (in mass per unit area per unit time), then the mass flux density for the ith component must be:

$$C_{ig}\rho_g \vec{v}_g + C_{io}\rho_o \vec{v}_o + C_{iw}\rho_w \vec{v}_w \tag{1-77}$$

The mass of component i per unit bulk volume of porous medium is:

$$\phi(C_{ig}\rho_g S_g + C_{io}\rho_o S_o + C_{iw}\rho_w S_w) \tag{1-78}$$

Further, we denote by q_i the mass rate of injection of component i per unit volume. Then, for each component, we can write the conservation equation as:

$$-\nabla \cdot [\alpha(C_{ig}\rho_g \vec{v}_g + C_{io}\rho_o \vec{v}_o + C_{iw}\rho_w \vec{v}_w)] + \alpha q_i$$

$$= \alpha \frac{\partial}{\partial t}[\phi(C_{ig}\rho_g S_g + C_{io}\rho_o S_o + C_{iw}\rho_w S_w)] \tag{1-79}$$

Differential equation

Darcy's law, as written previously for a three-phase system (eqs. (1-74)), still holds. Substitution of eq. (1-74) into (1-79) gives:

$$\nabla \cdot \left[\frac{\alpha C_{ig}\rho_g K k_{rg}}{\mu_g}(\nabla p_g - \rho_g g \nabla D) + \frac{\alpha C_{io}\rho_o K k_{ro}}{\mu_o}(\nabla p_o - \rho_o g \nabla D) \right.$$

$$\left. + \frac{\alpha C_{iw}\rho_w K k_{rw}}{\mu_w}(\nabla p_w - \rho_w g \nabla D) \right] + \alpha q_i$$

$$= \alpha \frac{\partial}{\partial t}[\phi(C_{ig}\rho_g S_g + C_{io}\rho_o S_o + C_{iw}\rho_w S_w)] \tag{1-80}$$

Auxiliary relations

While there are N of these differential equations, there are many more dependent variables. These are listed in Table I, and they total $3N + 15$.

TABLE I

Listing of all dependent variables

Variables	Number
C_{ig}	N
C_{io}	N
C_{iw}	N
p_g, p_o, p_w	3
S_g, S_o, S_w	3
ρ_g, ρ_o, ρ_w	3
μ_g, μ_o, μ_w	3
k_{rg}, k_{ro}, k_{rw}	3
	Total = $3N + 15$

In order to determine a solution to this system, we will need $3N + 15$ independent relations, differential, functional, or algebraic. In addition to the differential equations, we have the following functional or algebraic relations:

$$S_g + S_o + S_w = 1$$

Further, in each phase, the mass fractions must add up to 1.

$$\sum_{i=1}^{N} C_{ig} = \sum_{i=1}^{N} C_{io} = \sum_{i=1}^{N} C_{iw} = 1 \qquad (1\text{-}81)$$

Densities and viscosities are functions of the phase pressures and compositions:

$$\rho_g = f_1(p_g, C_{ig}) \qquad (1\text{-}82a)$$
$$\rho_o = f_2(p_o, C_{io}) \qquad (1\text{-}82b)$$
$$\rho_w = f_3(p_w, C_{iw}) \qquad (1\text{-}82c)$$
$$\mu_g = f_4(p_g, C_{ig}) \qquad (1\text{-}83a)$$
$$\mu_o = f_5(p_o, C_{io}) \qquad (1\text{-}83b)$$
$$\mu_w = f_6(p_w, C_{iw}) \qquad (1\text{-}83c)$$

Relative permeabilities are functions of saturation:

$$k_{rg} = f_7(S_g, S_o, S_w) \qquad (1\text{-}84a)$$
$$k_{ro} = f_8(S_g, S_o, S_w) \qquad (1\text{-}84b)$$
$$k_{rw} = f_9(S_g, S_o, S_w) \qquad (1\text{-}84c)$$

There are only two independent capillary pressure relationships:

$$p_g - p_o = p_{c_{go}}(S_g, S_o, S_w) \tag{1-85a}$$

$$p_o - p_w = p_{c_{ow}}(S_g, S_o, S_w) \tag{1-85b}$$

Finally, for each pair of phases, there is a distribution constant for each component; the distribution constant will be a function of pressure, temperature, and composition. Thus:

$$\frac{C_{ig}}{C_{io}} = K_{igo}(T, p_g, p_o, C_{ig}, C_{io}) \tag{1-86a}$$

$$\frac{C_{ig}}{C_{iw}} = K_{igw}(T, p_g, p_w, C_{ig}, C_{iw}) \tag{1-86b}$$

$$\frac{C_{io}}{C_{iw}} = K_{iow} = \frac{K_{igw}}{K_{igo}} \tag{1-86c}$$

However, the third equation is not independent of the first two; hence there are $2N$ relations for the phase equilibria.

Table II summarizes the number of relations; we see that there are indeed as many relations as there are dependent variables.

TABLE II

Listing of relations

Relation	Equations	Number
Partial differential equations	(1-80)	N
ΣS	(1-72)	1
ΣC_i	(1-81)	3
ρ	(1-82)	3
μ	(1-83)	3
k_r	(1-84)	3
Capillarity	(1-85)	2
Phase equilibrium	(1-86)	$2N$
	Total =	$3N + 15$

The black-oil model

Simplified, two-component, hydrocarbon system

The previous compositional model, while rigorous, is extremely complex to set up and solve. Such a model may be necessary for some highly volatile oil systems. However, for low-volatility oil systems, consisting mainly of methane and heavy components, a simplified "black oil", or two-component, model for describing the hydrocarbon equilibrium can be used, using data from a conventional differential vaporization test on the reservoir oil sample.

In the simplified model it is assumed that no mass transfer occurs between the water phase and the other two phases. In the hydrocarbon (oil-gas) system, only two components are considered. The "oil" component (also called stock-tank oil) is the residual liquid at atmospheric pressure left after a differential vaporization, while the "gas" component is the remaining fluid.

In order to reduce confusion, we need to distinguish carefully between gas component and gas phase, and between oil component and oil phase. Let us use capital letter subscripts to identify components and lower-case letter subscripts to identify phases. Further, we shall use the subscript S to indicate standard conditions.

Consider a sample of reservoir oil containing weights W_O of oil component and W_G of gas component. Let ρ_{OS} be the density of the oil component and ρ_{GS} the density of the gas component, both measured at standard conditions. Gas solubility, R_{so} (also called dissolved gas-oil ratio), is defined as the volume of gas (measured at standard conditions) dissolved at a given pressure and reservoir temperature in a unit volume of stock-tank oil. That is:

$$R_{so}(p, T) = V_{GS}/V_{OS} \tag{1-87}$$

Since:

$$V_{GS} = W_G/\rho_{GS} \tag{1-88}$$

and:

$$V_{OS} = W_O/\rho_{OS} \tag{1-89}$$

then:

$$R_{so} = W_G \rho_{OS}/(W_O \rho_{GS}) \tag{1-90}$$

Now, the volume of the oil phase at reservoir temperature and pressure is not V_{OS}, but somewhat larger, since the dissolved gas causes some swelling of the hydrocarbon liquid. The formation volume factor for oil, B_o, is defined as the ratio of the volume of oil plus its dissolved gas (measured at reservoir conditions) to the volume of the oil component measured at standard conditions. Thus:

$$B_o(p, T) = V_o(p, T)/V_{OS} \tag{1-91}$$

But:

$$V_o(p, T) = (W_O + W_G)/\rho_o \tag{1-92}$$

Combining eqs. (1-89), (1-91), and (1-92) yields:

$$B_o = (W_O + W_G)\rho_{OS}/(W_O \rho_o) \tag{1-93}$$

We can now determine the mass fraction of the two components in the oil phase. From eqs. (1-90) and (1-93) we obtain:

$$C_{Go} = W_G/(W_O + W_G) = R_{so}\rho_{GS}/(B_o \rho_o) \tag{1-94}$$

and from (1-93) we obtain:

$$C_{Oo} = W_O/(W_O + W_G) = \rho_{OS}/(B_o\rho_o) \tag{1-95}$$

The gas formation volume factor, B_g, is the ratio of the volume of free gas (all of which is gas component), measured at reservoir conditions, to the volume of the same gas measured at standard conditions. Thus:

$$B_g(p, T) = V_g(p, T)/V_{GS} \tag{1-96}$$

Let $W_g = W_G$ be the weight of free gas. Since $V_g = W_G/\rho_g$ and $V_{GS} = W_G/\rho_{GS}$, then:

$$B_g = \rho_{GS}/\rho_g \tag{1-97}$$

Finally, for completeness (and consistency), we define the water formation volume factor, B_w, in the same manner:

$$B_w = \rho_{WS}/\rho_w \tag{1-98}$$

Differential equations for black-oil model

To make use of the differential equations (1-80) previously derived for the compositional model, we need to define all the mass fractions, C_{ig}, C_{io}, and C_{iw}. Since the gas phase is all gas component, we have:

$$C_{Gg} = 1, \quad C_{Og} = 0, \quad C_{Wg} = 0$$

Similarly, the water phase is all water; hence

$$C_{Gw} = 0, \quad C_{Ow} = 0, \quad C_{Ww} = 1$$

In the oil phase, we already have obtained C_{Go} and C_{Oo} (eqs. (1-94) and (1-95)). In addition:

$$C_{Wo} = 0$$

Substitution of these mass fractions into eq. (1-80) gives, for the gas component $(i = G)$:

$$\nabla \cdot \left[\frac{\alpha \rho_g K k_{rg}}{\mu_g} (\nabla p_g - \rho_g g \nabla D) + \frac{\alpha R_{so} \rho_{GS} \rho_o K k_{ro}}{B_o \rho_o \mu_o} (\nabla p_o - \rho_o g \nabla D) \right]$$

$$+ \alpha q_G = \alpha \frac{\partial}{\partial t} \left[\phi \left(\rho_g S_g + \frac{R_{so} \rho_{GS} \rho_o S_o}{B_o \rho_o} \right) \right] \tag{1-99}$$

for the oil component $(i = O)$:

$$\nabla \cdot \left[\frac{\alpha \rho_{OS} \rho_o K k_{ro}}{B_o \rho_o \mu_o} (\nabla p_o - \rho_o g \nabla D) \right] + \alpha q_O = \alpha \frac{\partial}{\partial t} \left(\frac{\phi \rho_{OS} \rho_o S_o}{B_o \rho_o} \right) \tag{1-100}$$

and for the water component $(i = W)$:

$$\nabla \cdot \left[\frac{\alpha \rho_w K k_{rw}}{\mu_w} (\nabla p_w - \rho_w g \nabla D) \right] + \alpha q_w = \alpha \frac{\partial}{\partial t} (\phi \rho_w S_w) \qquad (1\text{-}101)$$

From eqs. (1-97) and (1-98) we have:

$$\rho_g = \rho_{GS}/B_g \qquad (1\text{-}102)$$

$$\rho_w = \rho_{WS}/B_w \qquad (1\text{-}103)$$

By adding eqs. (1-94) and (1-95) and noting that $C_{Go} + C_{Oo} = 1$, we obtain:

$$\rho_o = (R_{so}\rho_{GS} + \rho_{OS})/B_o \qquad (1\text{-}104)$$

By substitution for these densities into eqs. (1-99), (1-100), and (1-101), and dividing, respectively, by ρ_{GS}, ρ_{OS}, and ρ_{WS}, we obtain the final differential equations for the gas component:

$$\nabla \cdot \left[\frac{\alpha K k_{rg}}{B_g \mu_g} (\nabla p_g - \rho_g g \nabla D) + \frac{\alpha R_{so} K k_{ro}}{B_o \mu_o} (\nabla p_o - \rho_o g \nabla D) \right] + \frac{\alpha q_G}{\rho_{GS}}$$

$$= \alpha \frac{\partial}{\partial t} \left[\phi \left(\frac{S_g}{B_g} + \frac{R_{so} S_o}{B_o} \right) \right] \qquad (1\text{-}105)$$

for the oil component:

$$\nabla \cdot \left[\frac{\alpha K k_{ro}}{B_o \mu_o} (\nabla p_o - \rho_o g \nabla D) \right] + \frac{\alpha q_O}{\rho_{OS}} = \alpha \frac{\partial}{\partial t} \left(\frac{\phi S_o}{B_o} \right) \qquad (1\text{-}106)$$

and the water component:

$$\nabla \cdot \left[\frac{\alpha K k_{rw}}{B_w \mu_w} (\nabla p_w - \rho_w g \nabla D) \right] + \frac{\alpha q_W}{\rho_{WS}} = \alpha \frac{\partial}{\partial t} \left(\frac{\phi S_w}{B_w} \right) \qquad (1\text{-}107)$$

Note that these equations do not represent mass balances, but rather, balances on "standard volumes".

Limited compositional model

As mentioned previously, the black-oil model is not suitable for dealing with a volatile oil system. With a more elaborate two-component hydrocarbon system, the effect of oil volatility can be included, however. Cook, Jacoby, and Ramesh (1974) propose a scheme for modifying the parameters of a two-component system to represent more complex compositional effects even when three or more components are normally required. Further, the solubility of gas in water can easily be included.

Two-component hydrocarbon system with volatility

In this model, we assume again that there is both an oil and a gas component, but we permit solubility of gas in both the oil and water phases and permit vaporization of oil into the gas phase. Thus the "oil" component can exist in both the oil and gas phases, the "gas" component can exist in all three phases, and the "water" component can exist only in the water phase.

The solubility of gas in oil is given again by the solution gas-oil ratio:

$$R_{so} = V_{GS}/V_{OS} \qquad (1\text{-}108)$$

the solubility of gas in water by the solution gas-water ratio:

$$R_{sw} = V_{GS}/V_{WS} \qquad (1\text{-}109)$$

and the volatility of oil in the gas by the ratio:

$$R_v = V_{OS}/V_{GS} \qquad (1\text{-}110)$$

The mass fractions of the two components in the oil phase are the same as for the black-oil model:

$$C_{Go} = R_{so}\rho_{GS}/(B_o\rho_o) \qquad (1\text{-}111)$$

$$C_{Oo} = \rho_{OS}/(B_o\rho_o) \qquad (1\text{-}112)$$

By analogy, the mass fractions of the two components in the gas phase are:

$$C_{Og} = R_v\rho_{OS}/(B_g\rho_g) \qquad (1\text{-}113)$$

$$C_{Gg} = \rho_{GS}/(B_g\rho_g) \qquad (1\text{-}114)$$

while the mass fractions in the water phase are:

$$C_{Gw} = R_{sw}\rho_{GS}/(B_w\rho_w) \qquad (1\text{-}115)$$

$$C_{Ww} = \rho_{WS}/(B_w\rho_w) \qquad (1\text{-}116)$$

In addition:

$$C_{Wg} = C_{Wo} = C_{Ow} = 0 \qquad (1\text{-}117)$$

By adding eqs. (1-111) and (1-112) and noting that $C_{Go} + C_{Oo} = 1$, we obtain:

$$\rho_o = (\rho_{OS} + R_{so}\rho_{GS})/B_o \qquad (1\text{-}118)$$

Similarly:

$$\rho_g = (\rho_{GS} + R_v\rho_{OS})/B_g \qquad (1\text{-}119)$$

and

$$\rho_w = (\rho_{WS} + R_{sw}\rho_{GS})/B_w \qquad (1\text{-}120)$$

Differential equations

Substitution of these mass fractions and densities into eq. (1-80) and division, respectively, by ρ_{GS}, ρ_{OS}, and ρ_{WS}, gives, for the gas component ($i = G$):

$$\nabla \cdot \left[\frac{\alpha K k_{rg}}{B_g \mu_g} (\nabla p_g - \rho_g g \nabla D) + \frac{\alpha R_{so} K k_{ro}}{B_o \mu_o} (\nabla p_o - \rho_o g \nabla D) \right.$$
$$\left. + \frac{\alpha R_{sw} K k_{rw}}{B_w \mu_w} (\nabla p_w - \rho_w g \nabla D) \right] + \frac{\alpha q_G}{\rho_{GS}} = \alpha \frac{\partial}{\partial t} \left[\phi \left(\frac{S_g}{B_g} + \frac{R_{so} S_o}{B_o} + \frac{R_{sw} S_w}{B_w} \right) \right] \quad (1\text{-}121)$$

for the oil component ($i = O$):

$$\nabla \cdot \left[\frac{\alpha K k_{ro}}{B_o \mu_o} (\nabla p_o - \rho_o g \nabla D) + \frac{\alpha R_v K k_{rg}}{B_g \mu_g} (\nabla p_g - \rho_g g \nabla D) \right] + \frac{\alpha q_O}{\rho_{OS}}$$
$$= \alpha \frac{\partial}{\partial t} \left[\phi \left(\frac{S_o}{B_o} + \frac{R_v S_g}{B_g} \right) \right] \quad (1\text{-}122)$$

and the water component ($i = W$):

$$\nabla \cdot \left[\frac{\alpha K k_{rw}}{B_w \mu_w} (\nabla p_w - \rho_w g \nabla D) \right] + \frac{\alpha q_W}{\rho_{WS}} = \alpha \frac{\partial}{\partial t} \left(\frac{\phi S_w}{B_w} \right) \quad (1\text{-}123)$$

SUMMARY

In this chapter we have derived, in steps of increasing complexity, the differential equations that describe three-dimensional, three-phase flow with mass transfer between phases. While the fully compositional model has not yet achieved routine application, reservoir simulators based on the black-oil model and, to some extent, on the limited compositional model, are in widespread use throughout the industry. For this reason, the development of the differential equations for these models has been presented in considerable detail.

A complete description of the numerical techniques used to solve the most complicated of the mathematical models is beyond the scope of this book. Most of the numerical problems associated with multiphase reservoir simulation can be introduced by discussing the numerical solution of simpler problems, as we shall do in Chapters 3, 4, and 5. In Chapter 6, we shall conclude with the numerical solution of two-phase flow problems (without mass transfer).

It is true that three-phase simulators that include mass transfer are much more complicated than the two-phase models we shall cover. For the most part, however, they involve extensions of techniques used for solving two-phase problems. Certainly, some special problems arise in connection with the more elaborate models, and their importance should not be minimized. Nevertheless, even as we restrict the scope of the remaining chapters to two-phase numerical models, the reader should obtain a fundamental understanding of the numerical methods used in modern reservoir simulators.

NOMENCLATURE

Units

The symbols listed below are defined in terms of SI base units, kg (kilogram), m (meter), and s (second), plus two derived units, N (newton = kg · m/s^2) for force and Pa (pascal = N/m^2 = kg/m · s^2) for pressure.

Symbols

A	=	cross-sectional area of one-dimensional reservoir [m^2]
B	=	formation volume factor
c	=	compressibility [Pa^{-1} = m · s^2/kg]
C	=	concentration (i.e., mass fraction) of a component
D	=	depth [m]
$\mathfrak{D}$	=	diffusivity [m^2/s]
f_w	=	ratio of wetting-phase to total mobility, i.e., $\lambda_w/(\lambda_w + \lambda_n)$
F_w	=	fraction of flowing stream which is the wetting phase
g	=	acceleration due to gravity [m/s^2]
G_w	=	function of saturation and gravity, defined by eq. (1-60) [m/s]
h_w	=	function of saturation involving mobility and capillarity, defined by eq. (1-56) [m^2/s]
H	=	thickness of two-dimensional reservoir [m]
k_r	=	relative permeability
K	=	absolute permeability [m^2]
K_i	=	distribution constant for ith component
L	=	length [m]
M	=	molecular weight of gas [kg/mol]
N	=	number of components
p	=	pressure [Pa]
p_c	=	capillary pressure [Pa]
q	=	mass rate of injection (or production, if negative) per unit volume of reservoir [kg/m^{3}s]
Q	=	mass rate of injection (or production, if negative) at a well [kg/s]
Q_t	=	total volumetric rate of injection (or production, if negative) per unit volume of reservoir, equals $(q_n/\rho_n) + (q_w/\rho_w)$ [s^{-1}]
$\hat{Q}$	=	volumetric flow rate [m^3/s]
R	=	gas constant [Pa · m^3/°K · mol]
R_{so}	=	solubility of gas in oil
R_{sw}	=	solubility of gas in water
R_v	=	volatility of oil into gas
S	=	saturation
t	=	time [s]
T	=	absolute temperature [°K]
u	=	general function
v	=	flow velocity [m/s]
$\vec{v}$	=	flow velocity vector [m/s]
V	=	volume [m^3]
W	=	weight or mass [kg]
x	=	distance [m]
y	=	distance [m]
z	=	distance [m]
α	=	geometric factor, defined by eq. (1-13)

α_d = angle of inclination for one-dimensional reservoir
λ = phase mobility [m²/Pa · s = m³s/kg]
μ = viscosity [Pa · s = N · s/m² = kg/m · s]
ρ = density [kg/m³]
ϕ = porosity = void volume per unit bulk volume of porous medium
Φ = potential [Pa]

Subscripts

avg	average of nonwetting- and wetting-phase quantity
d	refers to difference between nonwetting- and wetting-phase quantity
g	refers to gas phase
G	refers to gas component
i	refers to ith component
n	refers to nonwetting phase
o	refers to oil phase
O	refers to oil component
S	at standard conditions
t	refers to total of nonwetting- and wetting-phase quantity
w	refers to wetting phase, or to water phase
W	refers to water component
x	refers to x-direction
y	refers to y-direction
z	refers to z-direction

CHAPTER 2

ELEMENTARY FINITE DIFFERENCES

INTRODUCTION

The numerical solution of partial differential equations by finite differences refers to the process of replacing the partial derivatives by finite-difference quotients, and then obtaining solutions of the resulting system of algebraic equations. In this chapter we shall examine some first- and second-order difference quotients and determine orders of accuracy for them. We shall do this for functions of two space variables, x and y, and of time, t. Extension of the discussion to three-dimensional systems (and reduction to one-dimensional systems) should be obvious.

FIRST-DIFFERENCE QUOTIENTS

Consider a function of three independent variables $u(x, y, t)$. A first derivative can be defined as a limit in several ways:

$$\frac{\partial u}{\partial x}(x, y, t) = \lim_{\Delta x \to 0} \frac{u(x + \Delta x, y, t) - u(x, y, t)}{\Delta x} \tag{2-1}$$

$$\frac{\partial u}{\partial x}(x, y, t) = \lim_{\Delta x \to 0} \frac{u(x, y, t) - u(x - \Delta x, y, t)}{\Delta x} \tag{2-2}$$

$$\frac{\partial u}{\partial x}(x, y, t) = \lim_{\Delta x \to 0} \frac{u(x + \Delta x, y, t) - u(x - \Delta x, y, t)}{2\Delta x} \tag{2-3}$$

Now, if we replace a derivative by a difference quotient, we want to know how good an approximation it is. For this, we use Taylor's series with remainder. For example: Expand $U(x+\Delta x)$ about x.

$$u(x + \Delta x, y, t) = u(x, y, t) + \Delta x \frac{\partial u}{\partial x}(x, y, t) + \frac{\Delta x^2}{2} \frac{\partial^2 u}{\partial x^2}(x^*, y, t)$$

where $x \leq x^* \leq x + \Delta x$. In this case, the last term is a remainder that involves the second derivative of u evaluated somewhere in the interval between x and $x + \Delta x$. Solving for $\partial u/\partial x$:

$$f(x,t) = f(x)|_{x_0} + f'(x)|_{x_0}(x-x_0) + f''(x)|_{x_0}\frac{(x-x_0)^2}{2!}$$

$$\frac{\partial u}{\partial x}(x, y, t) = \frac{u(x + \Delta x, y, t) - u(x, y, t)}{\Delta x} - \frac{\Delta x}{2}\frac{\partial^2 u}{\partial x^2}(x^*, y, t) \tag{2-4}$$

The expression:
$$\frac{u(x + \Delta x, y, t) - u(x, y, t)}{\Delta x}$$

is called a forward-difference quotient. We say that it is a first-order replacement for the derivative, $\partial u/\partial x$, in that the error is first order in Δx or, to put it another way, that the error is of the order of Δx. More precisely, the notation:

$$\epsilon = \mathcal{O}(\Delta x) \tag{2-5}$$

(aloud, we say that ϵ is big "oh" of Δx) means that ϵ is, in absolute value, at most a constant multiple of Δx. That is, eq. (2-5) is true if, and only if:

$$|\epsilon| \leq A|\Delta x| \tag{2-6}$$

for some constant, A.

Similarly, we can expand $u(x - \Delta x, y, t)$ in a Taylor series:

$$u(x - \Delta x, y, t) = u(x, y, t) - \Delta x \frac{\partial u}{\partial x}(x, y, t) + \frac{\Delta x^2}{2}\frac{\partial^2 u}{\partial x^2}(x^{**}, y, t)$$

$$(x - \Delta x \leq x^{**} \leq x)$$

$$\frac{\partial u}{\partial x}(x, y, t) = \frac{u(x, y, t) - u(x - \Delta x, y, t)}{\Delta x} + \frac{\Delta x}{2}\frac{\partial^2 u}{\partial x^2}(x^{**}, y, t) \tag{2-7}$$

so that the backward-difference quotient of eq. (2-2) is also first order.

On the other hand, eq. (2-3) involves a centered-difference quotient. To find its error, we need to use Taylor series expansions that are carried one term further. For brevity, we shall write u for $u(x, y, t)$, $\partial u/\partial x$ for $\partial u/\partial x(x, y, t)$, etc., and state arguments only when they apply to points other than (x, y, t).

$$u(x + \Delta x, y, t) = u + \Delta x \frac{\partial u}{\partial x} + \frac{\Delta x^2}{2!}\frac{\partial^2 u}{\partial x^2} + \frac{\Delta x^3}{3!}\frac{\partial^3 u}{\partial x^3}(x', y, t)$$

$$u(x - \Delta x, y, t) = u - \Delta x \frac{\partial u}{\partial x} + \frac{\Delta x^2}{2!}\frac{\partial^2 u}{\partial x^2} - \frac{\Delta x^3}{3!}\frac{\partial^3 u}{\partial x^3}(x'', y, t)$$

$$x \leq x' \leq x + \Delta x, \quad x - \Delta x \leq x'' \leq x$$

Subtracting the two equations:

$$u(x + \Delta x, y, t) - u(x - \Delta x, y, t) = 2\Delta x \frac{\partial u}{\partial x} + \frac{\Delta x^3}{6}\frac{\partial^3 u}{\partial x^3}(x', y, t) +$$

$$+ \frac{\Delta x^3}{6} \frac{\partial^3 u}{\partial x^3}(x'', y, t)$$

$$u(x + \Delta x, y, t) - u(x - \Delta x, y, t) = 2\Delta x \frac{\partial u}{\partial x} + \frac{\Delta x^3}{3} \frac{\partial^3 u}{\partial x^3}(x''', y, t)$$

where $x - \Delta x \leq x''' \leq x + \Delta x$. Solving for $\partial u/\partial x$:

$$\frac{\partial u}{\partial x} = \frac{u(x + \Delta x, y, t) - u(x - \Delta x, y, t)}{2\Delta x} - \frac{\Delta x^2}{6} \frac{\partial^3 u}{\partial x^3}(x''', y, t) \qquad (2\text{-}8)$$

Thus we see that the centered difference is of higher order, since the error term is $\mathcal{O}(\Delta x^2)$ instead of $\mathcal{O}(\Delta x)$. Intuitively, it would always seem advisable to use the centered-difference approximation in preference to the other two. However, this is not always true, as we shall see later when we consider solution of parabolic equations. Which form is preferable frequently depends on the particular problem.

We shall also want a difference quotient to replace a first derivative evaluated halfway between x and $x + \Delta x$. From eq. (2-8) we can see that:

$$\frac{\partial u}{\partial x}\left(x + \frac{\Delta x}{2}, y, t\right) = \frac{u(x + \Delta x, y, t) - u(x, y, t)}{\Delta x} - \frac{\Delta x^2}{24} \frac{\partial^3 u}{\partial x^3}(x', y, t) \qquad (2\text{-}9)$$

SECOND-DIFFERENCE QUOTIENTS

We can obtain an approximation for the second derivative [about K] by expanding $u(x + \Delta x, y, t)$ and $u(x - \Delta x, y, t)$ in Taylor series to remainders involving $\partial^4 u/\partial x^4$, and then *adding* the two expansions. If we do this, we find that:

$$\left.\frac{\partial^2 u}{\partial x^2}\right|_K = \frac{u(x + \Delta x, y, t) - 2u(x, y, t) + u(x - \Delta x, y, t)}{\Delta x^2} - \frac{\Delta x^2}{12} \frac{\partial^4 u}{\partial x^4}(x''', y, t)$$

$$x - \Delta x \leq x''' \leq x + \Delta x \qquad (2\text{-}10)$$

This centered second-difference quotient, which is of second-order accuracy, arises so frequently that we find it useful to define an operator by it:

$$\Delta_x^2 u(x, y, t) = \frac{u(x + \Delta x, y, t) - 2u(x, y, t) + u(x - \Delta x, y, t)}{\Delta x^2} \qquad (2\text{-}11)$$

While there are other difference expressions for $\partial^2 u/\partial x^2$, this one is used consistently; hence none of the others will be considered.

We have noted in Chapter 1 the very frequent occurrence of the second derivative of the form:

$$\frac{\partial}{\partial x}\left(K[x, y] \frac{\partial u}{\partial x}\right) \qquad (2\text{-}12)$$

In addition, we shall frequently be wanting to evaluate both $\partial^2 u/\partial x^2$ and (2-12) using unequally spaced intervals. That is, we will want to consider using values of u at $x - \Delta x'$, x, and $x + \Delta x''$, with $\Delta x'$ and $\Delta x''$ not necessarily being equal. Both of these extensions can be obtained by approximating the second derivative by taking the first difference of two first differences:

$$\frac{\partial}{\partial x}\left(K\frac{\partial u}{\partial x}\right) \approx \frac{\left(K\frac{\partial u}{\partial x}\right)_{x+\frac{1}{2}\Delta x''} - \left(K\frac{\partial u}{\partial x}\right)_{x-\frac{1}{2}\Delta x'}}{(x+\frac{1}{2}\Delta x'') - (x-\frac{1}{2}\Delta x')}$$

with:

$$\left(K\frac{\partial u}{\partial x}\right)_{x+\frac{1}{2}\Delta x''} \approx K(x+\tfrac{1}{2}\Delta x'')\frac{u(x+\Delta x'') - u(x)}{\Delta x''}$$

$$\left(K\frac{\partial u}{\partial x}\right)_{x-\frac{1}{2}\Delta x'} \approx K(x-\tfrac{1}{2}\Delta x')\frac{u(x) - u(x-\Delta x')}{\Delta x'}$$

$\times$ — See p 36

Then:

$$\frac{\partial}{\partial x}\left(K\frac{\partial u}{\partial x}\right) \approx \frac{K(x+\tfrac{1}{2}\Delta x'')\dfrac{u(x+\Delta x'')-u(x)}{\Delta x''} - K(x-\tfrac{1}{2}\Delta x')\dfrac{u(x)-u(x-\Delta x')}{\Delta x'}}{\dfrac{\Delta x' + \Delta x''}{2}}$$

(2-13)

This difference quotient we shall refer to as the operator

$$\Delta_x(K\Delta_x u) \tag{2-14}$$

GRID SYSTEMS

For two space variables, we now consider the grid system with which we divide up the solution region in the x-y plane. The integer i is used as the index in the x-direction, and the integer j for the index in the y-direction. Thus x_i is the ith value of x, and y_j is the jth value of y. Double indexing is used to identify functions within the two-dimensional region. Thus:

$$u_{ij} = u(x_i, y_j)$$

We shall now proceed to examine in some detail two types of grids commonly used in reservoir work.

Block-centered grid

In one type of grid, the rectangle is divided into blocks, as in Fig. 5, and the point (x_i, y_j) is considered to be at the center of block (i, j). There are I blocks in the x-direction and J blocks in the y-direction.

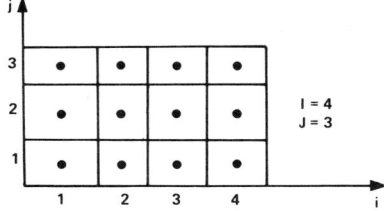

Fig. 5. Rectangle divided into blocks.

Further details concerning this grid are given in Fig. 6. We identify the coordinate $x_{i-\frac{1}{2}}$ with the left side of the block (i, j), and $x_{i+\frac{1}{2}}$ with the right side of the block. Similarly, $y_{j-\frac{1}{2}}$ is identified with the bottom of the block, and $y_{j+\frac{1}{2}}$ with the top. This type of grid is called a "block-centered" grid. We note that the grid is specified by the sequences $x_{\frac{1}{2}} = 0, x_{\frac{3}{2}}, x_{\frac{5}{2}}, \ldots, x_{I+\frac{1}{2}}$ and $y_{\frac{1}{2}} = 0, y_{\frac{3}{2}}, y_{\frac{5}{2}}, \ldots, y_{J+\frac{1}{2}}$. Also note that:

$$x_i = \tfrac{1}{2}(x_{i-\frac{1}{2}} + x_{i+\frac{1}{2}}) \quad \text{and} \quad y_j = \tfrac{1}{2}(y_{j-\frac{1}{2}} + y_{j+\frac{1}{2}}) \tag{2-15}$$

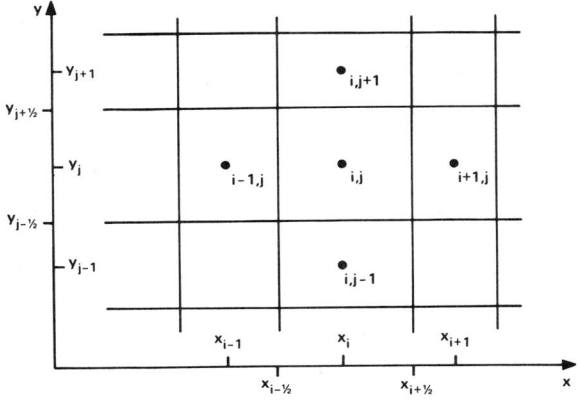

Fig. 6. Details of block-centered grid.

Point-centered grid

In the other type of grid, the rectangle is divided up by a grid system as shown in Fig. 7. The point (x_i, y_j), also called the point (i, j), is now considered to be at the *intersection* of each grid line. There are I grid lines in the x-direction and J grid lines in the y-direction.

We refer to this as a "point-centered" grid. Details are shown in Fig. 8. We can still maintain the notion of a block located about each point (i, j), as shown by the dashed lines in Figs. 7 and 8, but in this case the point is not

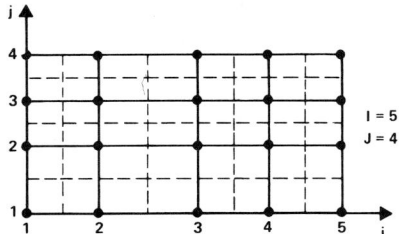

Fig. 7. Rectangle divided up by system of grid lines.

necessarily in the center of its block. Rather, the block boundaries are located midway between the points.

Here the grid is specified by the sequences $x_1 = 0, x_2, \ldots, x_I$, and $y_1 = 0, y_2, \ldots, y_J$, and:

$$x_{i+\frac{1}{2}} = \tfrac{1}{2}(x_i + x_{i+1}) \quad \text{and} \quad x_{j+\frac{1}{2}} = \tfrac{1}{2}(y_j + y_{j+1}) \qquad (2\text{-}16)$$

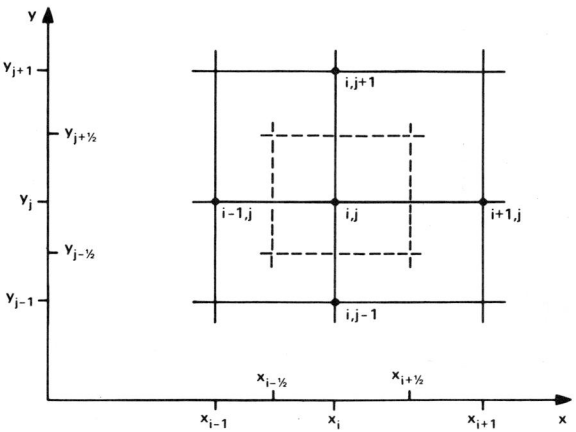

Fig. 8. Details of point-centered grid.

Comparison of the two grids

We shall see that the difference equations will be written the same way for either grid. There are, however, two significant differences between the two grid systems. First, when the spacing is nonuniform, the locations of the points and the block boundaries will not coincide. Second, the treatment at the boundaries of the computing rectangle is quite different.

For the block-centered grid, there are no points on the boundary of the computing rectangle, only blocks, and the difference representation for the boundary conditions $\partial u/\partial x = 0$ and $\partial u/\partial y = 0$ is:

$$U_{0,j} = U_{1,j}, \quad U_{I+1,j} = U_{I,j}, \quad 1 \leqslant j \leqslant J$$
$$U_{i,0} = U_{i,1}, \quad U_{i,J+1} = U_{i,J}, \quad 1 \leqslant i \leqslant I \tag{2-17}$$

On the other hand, for the point-centered grid, there *are* points on the boundary of the computing rectanlge, and the difference representations of the same boundary conditions are:

$$U_{0,j} = U_{2,j}, \quad U_{I+1,j} = U_{I-1,j} \quad 1 \leqslant j \leqslant J$$
$$U_{i,0} = U_{i,2}, \quad U_{i,J+1} = U_{i,J-1} \quad 1 \leqslant i \leqslant I \tag{2-18}$$

Equations (2-17) and (2-18) are known as reflection boundary conditions, and they utilize a set of fictitious points located either one half-interval or one interval outside of the solution rectangle.

The difference quotient (2-13) has the same appearance for both types of grids. The derivatives are approximated by:

$$\frac{\partial}{\partial x}\left(K\frac{\partial u}{\partial x}\right) \approx \frac{K_{i+\frac{1}{2},j}\dfrac{U_{i+1,j} - U_{ij}}{x_{i+1} - x_i} - K_{i-\frac{1}{2},j}\dfrac{U_{ij} - U_{i-1,j}}{x_i - x_{i-1}}}{x_{i+\frac{1}{2}} - x_{i-\frac{1}{2}}} \tag{2-19}$$

$$\frac{\partial}{\partial y}\left(K\frac{\partial u}{\partial y}\right) \approx \frac{K_{i,j+\frac{1}{2}}\dfrac{U_{i,j+1} - U_{ij}}{y_{j+1} - y_j} - K_{i,j-\frac{1}{2}}\dfrac{U_{ij} - U_{i,j-1}}{y_j - y_{j-1}}}{y_{j+\frac{1}{2}} - y_{j-\frac{1}{2}}} \tag{2-20}$$

TRUNCATION ERROR

Truncation error is that error incurred by replacing a differential equation by a difference equation. Due to this error, the exact solution (i.e., with no round-off error) of a difference equation differs from the solution of the corresponding differential equation. The term "truncation error" derives, of course, from the fact that the replacement of a derivative by a difference quotient is equivalent to using a *truncated* Taylor's series.

Suppose we wish to solve the heat equation

$$\frac{\partial^2 u}{\partial x^2} = \frac{\partial u}{\partial t} \tag{2-21}$$

by use of the difference equation

$$\frac{U_{i-1,n} - 2U_{in} + U_{i+1,n}}{\Delta x^2} = \frac{U_{i,n+1} - U_{in}}{\Delta t} \tag{2-22}$$

where i is the index in the x-direction and n is the index in the t-direction. Note that capital letters are used to indicate the solution to a difference equation, whereas we have been using lower-case letters to represent the solution to a differential equation.

A "local" truncation error of a finite-difference approximation can be defined by:

$$\epsilon_L = L_D U_{in} - (Lu)_{x_i, t_n} \qquad (2\text{-}23)$$

where ϵ_L = local truncation error, $L_D U$ = difference form, and Lu = differential form.

Consider eqs. (2-21) and (2-22). Then:

$$L_D U_{in} = \frac{U_{i-1,n} - 2U_{in} + U_{i+1,n}}{\Delta x^2} - \frac{U_{i,n+1} - U_{in}}{\Delta t}$$

$$Lu = \frac{\partial^2 u}{\partial x^2} - \frac{\partial u}{\partial t}$$

Substituting eq. (2-10) for $\partial^2 u/\partial x^2$ and rewriting (2-4) as:

$$\frac{\partial u}{\partial t} = \frac{u(x, t + \Delta t) - u(x, t)}{\Delta t} - \frac{\Delta t}{2} \frac{\partial^2 u}{\partial t^2}(x, t^*)$$

gives for ϵ_L:

$$\epsilon_L = \frac{\Delta x^2}{12} \frac{\partial^4 u}{\partial x^4}(x''', t) - \frac{\Delta t}{2} \frac{\partial^2 u}{\partial t^2}(x, t^*)$$

$$\epsilon_L = \mathcal{O}(\Delta x^2) + \mathcal{O}(\Delta t) \qquad (2\text{-}24)$$

If we had the solutions, we could define the "global" truncation error by:

$$\epsilon_{G,1} = \max_{i,n} |U_{in} - u(x_i, t_n)| \qquad (2\text{-}25)$$

Another global measure of error frequently used is:

$$\epsilon_{G,2} = \left[\frac{1}{IN} \sum_{i=1}^{I} \sum_{n=1}^{N} (U_{in} - u[x_i, t_n])^2 \right]^{\frac{1}{2}} \qquad (2\text{-}26)$$

Of course, in general, we don't have the solution, so it becomes one of the tasks of the numerical analyst to attempt to derive bounds for the global errors. This can be done with moderate difficulty for simple differential equations; as the equations become more complex, estimates of the bounds become more and more difficult to obtain. Usually, however, it turns out that the global errors have the same order of dependence on mesh size as the local error. Thus eq. (2-22) has a global error (Douglas, 1961, p. 8) that is also $\mathcal{O}(\Delta x^2) + \mathcal{O}(\Delta t)$. Hence the local truncation error, which is much easier to estimate, can at least be used as a guide to the order of convergence of the solution of the difference equation to the solution of the differential equation as the mesh size is decreased to zero.

As a practical matter, for problems involving any complexity at all, estimates of the error are best obtained by solving the difference equations with different mesh sizes, varying both Δx (and Δy and Δz) and Δt to find their effects on the solution. In many cases, practical values of Δx and Δt (wherein the computational work is not excessive) may be so large that the error does not appear to decrease as rapidly as predicted by the formulas for local truncation error. The reason for this is that expressions for the order of the error describe the *asymptotic* behavior as Δx and Δt approach zero and really say nothing about the behavior of the error for large mesh sizes. Consequently, we must content ourselves frequently with empirical error estimates obtained by running the same problem with several different mesh sizes. We would then run the remainder of the cases with the mesh size that balances the risks associated with the apparent error against the cost of using a smaller grid size.

NOMENCLATURE

Symbols

- I = number of blocks or grid lines in x-direction
- J = number of blocks or grid lines in y-direction
- K = absolute permeability [m^2]
- L = differential operator
- L_D = difference operator
- t = time [s]
- u = general function
- U = difference approximation for function u
- x = distance [m]
- y = distance [m]
- Δt = increment of t
- Δx = increment of x
- $\Delta_x^2 u$ = second-difference quotient for equal Δx, defined by eq. (2-11)
- $\Delta_x^2(K\Delta_x u)$ = second-difference quotient for variable K or unequal Δx, defined by eq. (2-13)
- ϵ = error

Subscripts

- G global
- i index in x-direction
- j index in y-direction
- L local
- n index in t-direction (i.e., time)

CHAPTER 3

NUMERICAL SOLUTION OF PARABOLIC PROBLEMS IN ONE DEPENDENT VARIABLE

THE FORWARD-DIFFERENCE EQUATION

Consider the simple parabolic problem, the heat-flow equation in one space dimension:

$$\frac{\partial^2 u}{\partial x^2} = \frac{\partial u}{\partial t} \tag{3-1}$$

with initial and boundary conditions:

$$u(x, 0) = f_0(x), \quad u(x_1, t) = g_1(t), \quad u(x_2, t) = g_2(t) \tag{3-2}$$

A numerical procedure for solving parabolic problems will generally consist of some method for obtaining the solution at time t_{n+1} from a solution already obtained at time t_n. Thus we can proceed from the initial condition (at $t = 0$) to a solution at $t = t_1 = \Delta t_0$; using this, we obtain a solution at $t_2 = \Delta t_0 + \Delta t_1$, and so on, to any t_n. Hence the solution "marches" through time.

The simplest difference analog to eq. (3-1) is obtained by replacing $\partial^2 u/\partial x^2$ by a second-difference evaluated at t_n and replacing $\partial u/\partial t$ by its forward-difference analog. This gives the centered-in-distance, forward-in-time difference equation (3-3), which is better known as the *forward-difference equation*:

$$\frac{U_{i+1,n} - 2U_{in} + U_{i-1,n}}{\Delta x^2} = \frac{U_{i,n+1} - U_{in}}{\Delta t}, \quad 1 \leq i \leq I-1 \tag{3-3}$$

which we can solve *explicitly* for $U_{i,n+1}$:

$$U_{i,n+1} = U_{in} + (\Delta t/\Delta x^2)(U_{i+1,n} - 2U_{in} + U_{i-1,n}) \tag{3-4}$$

with initial condition

$$U_{i,0} = f_0(x_i)$$

and boundary conditions

$$U_{0,n} = g_1(t_n)$$

$$U_{I,n} = g_2(t_n)$$

The use of marching procedures brings about the *stability* problem. This might best be introduced by a simple numerical example. Let us solve eq. (3-1) in the interval $0 < x < 1$ with initial and boundary conditions

$$f_0(x) = 1 \quad \text{and} \quad g_1(t) = g_2(t) = 0$$

and use eq. (3-4) with $\Delta x = \frac{1}{4}$. We will try several values of Δt, i.e., $\frac{1}{64}$, $\frac{1}{32}$, and $\frac{1}{16}$. The ratio $\Delta t/\Delta x^2$ will have the values $\frac{1}{4}$, $\frac{1}{2}$, and 1, respectively. The numerical solutions up to $t = 0.3125$ are given in Table III. We see that, for $\Delta t/\Delta x^2 = 1$, the solution diverges and becomes quite meaningless.

Clearly, then, in considering various difference analogs to a differential equation, we must attempt not only to evaluate truncation error, but also to examine their stability, and to determine the effect of time-step size on stability. One of the most useful ways to examine stability is by means of harmonic analysis.

STABILITY BY HARMONIC ANALYSIS (THE VON NEUMANN CRITERION)

We call a computation scheme *stable* if the effect of an error (or perturbation) made in one stage of the computation is not propagated into larger errors in latter stages of the computation. In other words, local errors are not magnified by further computation. Stability analysis need not concern itself with the source of error, whether it be due to round-off error, truncation error, or some other cause.

A difference equation is tested for stability by substituting into it perturbed values of the solution. For example, if $U_{i,n}$ is a solution to the difference equation, we assume that its perturbation,

$$U_{i,n} + \epsilon_{i,n}$$

also satisfies the difference equation, and we examine the possible growth of the perturbation, $\epsilon_{i,n}$. Specifically, if $U_{i,n}$ satisfies eq. (3-3), then:

$$\frac{(U_{i+1,n} + \epsilon_{i+1,n}) - 2(U_{i,n} + \epsilon_{i,n}) + (U_{i-1,n} + \epsilon_{i-1,n})}{\Delta x^2}$$
$$= \frac{(U_{i,n+1} + \epsilon_{i,n+1}) - (U_{i,n} + \epsilon_{i,n})}{\Delta t} \quad (3\text{-}5)$$

Subtracting eq. (3-3) from (3-5) gives:

$$\frac{\epsilon_{i+1,n} - 2\epsilon_{i,n} + \epsilon_{i-1,n}}{\Delta x^2} = \frac{\epsilon_{i,n+1} - \epsilon_{i,n}}{\Delta t} \quad (3\text{-}6)$$

We see that the error equation (3-6) has exactly the same form as the original difference equation (3-3). It will generally be true that a difference equation

TABLE III

Solutions of explicit difference analog of heat-flow equation

$\Delta t/\Delta x^2$	n	t	$U_{i,n}$				
			$i=0$	$i=1$	$i=2$	$i=3$	$i=4$
0.25	0	.000000	.00000	1.00000	1.00000	1.00000	.00000
	1	.015625	.00000	.75000	1.00000	.75000	.00000
	2	.031250	.00000	.62500	.87500	.62500	.00000
	3	.046875	.00000	.53125	.75000	.53125	.00000
	4	.062500	.00000	.45312	.64062	.45312	.00000
	5	.078125	.00000	.38671	.54687	.38671	.00000
	6	.093750	.00000	.33007	.46679	.33007	.00000
	7	.109375	.00000	.28173	.39843	.28173	.00000
	8	.125000	.00000	.24047	.34008	.24047	.00000
	9	.140625	.00000	.20526	.29028	.20526	.00000
	10	.156250	.00000	.17520	.24777	.17520	.00000
	11	.171875	.00000	.14954	.21148	.14954	.00000
	12	.187500	.00000	.12764	.18051	.12764	.00000
	13	.203125	.00000	.10895	.15407	.10895	.00000
	14	.218750	.00000	.09299	.13151	.09299	.00000
	15	.234375	.00000	.07937	.11225	.07937	.00000
	16	.250000	.00000	.06775	.09581	.06775	.00000
	17	.265625	.00000	.05782	.08178	.05782	.00000
	18	.281250	.00000	.04936	.06980	.04936	.00000
	19	.296875	.00000	.04213	.05958	.04213	.00000
	20	.312500	.00000	.03596	.05085	.03596	.00000
0.5	0	.000000	.00000	1.00000	1.00000	1.00000	.00000
	1	.031250	.00000	.50000	1.00000	.50000	.00000
	2	.062500	.00000	.50000	.50000	.50000	.00000
	3	.093750	.00000	.25000	.50000	.25000	.00000
	4	.125000	.00000	.25000	.25000	.25000	.00000
	5	.156250	.00000	.12500	.25000	.12500	.00000
	6	.187500	.00000	.12500	.12500	.12500	.00000
	7	.218750	.00000	.06250	.12500	.06250	.00000
	8	.250000	.00000	.06250	.06250	.06250	.00000
	9	.281250	.00000	.03125	.06250	.03125	.00000
	10	.312500	.00000	.03125	.03125	.03125	.00000
1.0	0	.000000	.00000	1.00000	1.00000	1.00000	.00000
	1	.062500	.00000	0.00000	1.00000	0.00000	.00000
	2	.125000	.00000	1.00000	−1.00000	1.00000	.00000
	3	.187500	.00000	−2.00000	3.00000	−2.00000	.00000
	4	.250000	.00000	5.00000	−7.00000	5.00000	.00000
	5	.312500	.00000	−12.00000	17.00000	−12.00000	.00000

and its error equation will be identical when the difference equation is linear and homogeneous. (Nonhomogeneous terms in the difference equation arising from flow terms or certain boundary conditions will, however, be absent from the error equation.)

The well-known von Neumann stability analysis, described in some detail by Richtmyer (1957) and Ames (1969), consists in expanding the error in a Fourier series of the form:

$$\epsilon_{i,n} = \sum_p \gamma_p^n \exp(ipx_i) \qquad (3\text{-}7)$$

substituting the series into the error equation (or, for that matter, directly into the homogeneous linear difference equation) and solving for the amplification factor:

$$\gamma_p = \gamma_p^{n+1}/\gamma_p^n$$

for each component. The von Neumann criterion for stability is that the modulus of the amplification factor must be less than or equal to one for all the components.

The test for stability can be simplified somewhat by dropping the subscript p and by taking

$$x_i = i \cdot \Delta x$$

We assume that the homogeneous difference equation has a solution of the form:

$$\epsilon_{i,n} = \gamma^n \exp(ipi\Delta x) \qquad (3\text{-}8)$$

Substitution of this assumed solution directly into the difference equation and cancellation of the resulting common factor, $\gamma^n \exp(ipi\Delta x)$, leads to an equation that must be satisfied by the parameters γ, p, Δx, and Δt. The von Neumann criterion for stability is the requirement that it be impossible to satisfy this equation by any real or complex value of γ for which $|\gamma| > 1$.

This criterion ignores the boundary conditions, and hence would be truly valid only for pure initial value problems (i.e., where the domain of the initial value and of the solution is the entire x-axis, $-\infty < x < \infty$). However, it has been our experience that properly posed boundary conditions have little effect on stability and that the von Neumann condition is an extremely valuable guide to stability for the practical computer.

Let us apply the criterion to the forward-difference equation (3-3). Substitution of eq. (3-8) gives:

$$[\gamma^n \exp\{ip(i+1)\Delta x\} - 2\gamma^n \exp\{ipi\Delta x\} + \gamma^n \exp\{ip(i-1)\Delta x\}]/\Delta x^2$$
$$= [\gamma^{n+1} \exp\{ipi\Delta x\} - \gamma^n \exp\{ipi\Delta x\}]/\Delta t$$

or:

$$[\gamma^n \exp(ip\Delta x) - 2\gamma^n + \gamma^n \exp(-ip\Delta x)]/\Delta x^2 = [\gamma^{n+1} - \gamma^n]/\Delta t \qquad (3\text{-}9)$$

But:

$$\exp(ip\Delta x) + \exp(-ip\Delta x) - 2 = 2\cosh(ip\Delta x) - 2$$
$$= 2\cos(p\Delta x) - 2$$
$$= -4\sin^2(p\Delta x/2) \tag{3-10}$$

By substituting eq. (3-10) into (3-9), we obtain:

$$\gamma^n[-4\sin^2(p\Delta x/2)]/\Delta x^2 = [\gamma^{n+1} - \gamma^n]/\Delta t$$
$$\gamma^{n+1} = \gamma^n - \gamma^n(4\Delta t/\Delta x^2)\sin^2(p\Delta x/2) \tag{3-11}$$

Dividing by γ^n, we obtain the "stability ratio", γ:

$$\gamma = 1 - (4\Delta t/\Delta x^2)\sin^2(p\Delta x/2) \tag{3-12}$$

The von Neumann criterion for stability is satisfied if:

$$-1 \leqslant 1 - (4\Delta t/\Delta x^2)\sin^2(p\Delta x/2) \leqslant 1$$

The right-hand inequality is satisfied trivially, since Δt and Δx^2 are positive. The left-hand inequality is:

$$-2 \leqslant -(4\Delta t/\Delta x^2)\sin^2(p\Delta x/2)$$

$$\Delta t/\Delta x^2 \leqslant \frac{1}{2\sin^2(p\Delta x/2)} \tag{3-13}$$

While the maximum value of the denominator is slightly less than 2 (how much less depends on I, the number of points in the x-direction), as a practical matter we can ignore the difference and merely require that:

$$\Delta t/\Delta x^2 \leqslant \tfrac{1}{2} \tag{3-14}$$

in order to obtain stability. This result is in agreement with our example calculations (Table III). The forward-difference equation is called conditionally stable.

IMPLICIT DIFFERENCE EQUATIONS

The backward-difference equation

The restriction on time step that is inherent in the forward-difference equation (an explicit equation) can be removed by evaluating the second derivative at time t_{n+1}, instead of at t_n. Thus, we write:

$$\frac{U_{i+1,n+1} - 2U_{i,n+1} + U_{i-1,n+1}}{\Delta x^2} = \frac{U_{i,n+1} - U_{in}}{\Delta t} \tag{3-15}$$

As we are going from solutions $U_{i,n}$ to $U_{i,n+1}$, each equation (3-15) involves three unknown values of $U_{i,n+1}$. Thus eq. (3-15) defines $U_{i,n+1}$

implicitly, and a tridiagonal system of simultaneous equations must be solved at each time step. For this, the tridiagonal algorithm discussed in the next section provides an efficient method of solution, so that the work involved in the solution of eq. (3-15) is little more than that involved in carrying out the explicit calculation of the forward-difference equation.

Let us perform a von Neumann stability analysis on eq. (3-15). Substituting eq. (3-8) gives:

$$[\gamma^{n+1} \exp\{ip(i+1)\Delta x\} - 2\gamma^{n+1} \exp\{ipi\Delta x\} + \gamma^{n+1} \exp\{ip(i-1)\Delta x\}]/\Delta x^2$$
$$= [\gamma^{n+1} \exp\{ipi\Delta x\} - \gamma^n \exp\{ipi\Delta x\}]/\Delta t$$

$$\gamma^{n+1}[\exp(ip\Delta x) - 2 + \exp(-ip\Delta x)]/\Delta x^2 = [\gamma^{n+1} - \gamma^n]/\Delta t$$

$$\gamma^{n+1}[-4\sin^2(p\Delta x/2)]/\Delta x^2 = [\gamma^{n+1} - \gamma^n]/\Delta t$$

$$\gamma^{n+1} = \frac{\gamma^n}{1 + (4\Delta t/\Delta x^2)\sin^2(p\Delta x/2)} \tag{3-16}$$

$$\gamma = \frac{1}{1 + (4\Delta t/\Delta x^2)\sin^2(p\Delta x/2)} \tag{3-17}$$

For any choice of p, Δt, or Δx, $\gamma < 1$; hence eq. (3-15) is unconditionally stable.

The tridiagonal algorithm

Let us consider the general tridiagonal system of equations shown in Fig. 9. A straightforward Gaussian elimination scheme consists first of solving for x_1 and then eliminating it from the second equation (note that x_1 does not appear in any other equations):

$$x_1 = \frac{d_1}{b_1} - \frac{c_1 x_2}{b_1}$$

Let us define:
$$\gamma_1 = d_1/b_1, \quad w_1 = c_1/b_1$$

$$\begin{aligned}
b_1 x_1 + c_1 x_2 &= d_1 \\
a_2 x_1 + b_2 x_2 + c_2 x_3 &= d_2 \\
a_3 x_2 + b_3 x_3 + c_3 x_4 &= d_3 \\
&\vdots \\
a_I x_{I-1} + b_I x_I &= d_I
\end{aligned}$$

Fig. 9. Tridiagonal system of equations.

Then:

$$x_1 = \gamma_1 - w_1 x_2$$

After substituting for x_1, the second equation becomes:

$$a_2(\gamma_1 - w_1 x_2) + b_2 x_2 + c_2 x_3 = d_2$$

$$(b_2 - a_2 w_1)x_2 + c_2 x_3 = d_2 - a_2 \gamma_1$$

Define $\beta_2 = b_2 - a_2 w_1$. Then:

$$x_2 = \frac{d_2 - a_2 \gamma_1}{\beta_2} - \frac{c_2 x_3}{\beta_2}$$

Define:

$$\gamma_2 = \frac{d_2 - a_2 \gamma_1}{\beta_2}, \quad w_2 = \frac{c_2}{\beta_2}$$

Then:

$$x_2 = \gamma_2 - w_2 x_3$$

If we continue this process, up to $i = I - 1$, with:

$$\beta_i = b_i - a_i w_{i-1}, \quad \gamma_i = \frac{d_i - a_i \gamma_{i-1}}{\beta_i}, \quad \text{and} \quad w_i = \frac{c_i}{\beta_i}$$

we obtain:

$$x_{I-1} = \gamma_{I-1} - w_{I-1} x_I$$

Substitution into the last equation gives:

$$a_I(\gamma_{I-1} - w_{I-1} x_I) + b_I x_I = d_I$$

$$(b_I - a_I w_{I-1})x_I = d_I - a_I \gamma_{I-1}$$

With the final definitions of:

$$\beta_I = b_I - a_I w_{I-1} \quad \text{and} \quad \gamma_I = \frac{d_I - a_I \gamma_{I-1}}{\beta_I}$$

then:

$$x_I = \gamma_I$$

We can now put together the entire tridiagonal algorithm. In order of increasing i, we calculate (forward solution):

$$\beta_1 = b_1 \tag{3-18a}$$

$$w_i = c_i/\beta_i \quad 1 \leq i \leq I-1 \tag{3-19}$$

$$\beta_i = b_i - a_i w_{i-1} \quad 2 \leq i \leq I \tag{3-18b}$$

$$\gamma_1 = d_1/\beta_1$$

$$\gamma_i = \frac{d_i - a_i \gamma_{i-1}}{\beta_i} \quad 2 \leq i \leq I \tag{3-20}$$

Then, in order of decreasing i, we calculate (back solution):

$$x_I = \gamma_I$$

$$x_i = \gamma_i - w_i x_{i+1} \quad 1 \leq i \leq I-1 \tag{3-21}$$

This algorithm is of great importance. Not only is it useful for solution of one-dimensional difference equations, such as (3-15), but it forms the basis of several methods for solving two-dimensional (and even three-dimensional) problems, as we shall see later. An important property is that the computing labor is directly proportional to the number of equations, I, compared with the $N^3/2$ labor required to solve N equations each containing N unknowns.

The Crank-Nicolson difference equation

Another implicit difference analog for the heat equation (3-1) can be obtained by replacing $\partial^2 u/\partial t^2$ by the average of the second-difference quotients at both the old and new time levels, t_n, and t_{n+1}. This is called the Crank-Nicolson equation.

$$\frac{1}{2}\left[\frac{U_{i+1,n} - 2U_{in} + U_{i-1,n}}{\Delta x^2} + \frac{U_{i+1,n+1} - 2U_{i,n+1} + U_{i-1,n+1}}{\Delta x^2}\right]$$

$$= \frac{U_{i,n+1} - U_{i,n}}{\Delta t} \tag{3-22}$$

It also gives rise to a tridiagonal system of equations. By harmonic analysis, we can show that the stability ratio is given by:

$$\gamma = \frac{1 - (2\Delta t/\Delta x^2)\sin^2(p\Delta x/2)}{1 + (2\Delta t/\Delta x^2)\sin^2(p\Delta x/2)} \tag{3-23}$$

There are no values of $\Delta t, \Delta x$, and p for which the magnitude of γ exceeds 1, so the Crank-Nicolson equation is also unconditionally stable.

While the Crank-Nicolson equation appears slightly more complicated to calculate than the backward-difference equation (3-15), its appeal lies in the higher order accuracy indicated by the order of the error. We have already seen in Chapter 2 that the forward-difference equation has a local truncation error:

F-D: $\epsilon_L = \mathcal{O}(\Delta x^2) + \mathcal{O}(\Delta t)$ \hfill (3-24)

By using the same method, we can easily show that the backward-difference equation (3-15) has the same order of accuracy:

B-D: $\quad \epsilon_L = \mathcal{O}(\Delta x^2) + \mathcal{O}(\Delta t)$ \hfill (3-25)

For the Crank-Nicolson equation (3-22), we can show that:

C-N: $\quad \epsilon_L = \mathcal{O}(\Delta x^2) + \mathcal{O}(\Delta t^2)$ \hfill (3-26)

OTHER EXPLICIT DIFFERENCE EQUATIONS

A time-centered explicit equation

There appears to be a simpler way than the Crank-Nicolson method to obtain a second-order correct estimate for $\partial u/\partial t$ (Douglas, 1961, p. 24). Consider the difference equation:

$$\frac{U_{i+1,n} - 2U_{in} + U_{i-1,n}}{\Delta x^2} = \frac{U_{i,n+1} - U_{i,n-1}}{2\Delta t} \tag{3-27}$$

This is an example of a three-level difference equation, where solutions at t_{n-1} and t_n are used to produce a solution at t_{n+1} explicitly. The first application of this equation must be to obtain $U_{i,2}$ from $U_{i,0}$ and $U_{i,1}$; some other means would have to be used to obtain $U_{i,1}$. Let us forget this difficulty for the moment and apply the von Neumann stability analysis:

$[\gamma^n \exp\{ip(i+1)\Delta x\} - 2\gamma^n \exp\{ipi\Delta x\} + \gamma^n \exp\{ip(i-1)\Delta x\}]/\Delta x^2$

$\quad = [\gamma^{n+1} \exp\{ipi\Delta x\} - \gamma^{n-1} \exp\{ipi\Delta x\}]/2\Delta t$

$\gamma^n[-8(\Delta t/\Delta x^2)\sin^2(p\Delta x/2)] = \gamma^{n+1} - \gamma^{n-1}$

$\gamma^2 + 8\gamma r \sin^2(p\Delta x/2) - 1 = 0$

where $r = \Delta t/\Delta x^2$.

The appearance of the quadratic expression for γ arises from the fact that eq. (3-27) is a three-level difference equation. We can then obtain:

$\gamma = -4r\sin^2(p\Delta x/2) \pm [16r^2 \sin^4(p\Delta x/2) + 1]^{\frac{1}{2}}$

Let $R = 4r\sin^2(p\Delta x/2)$. For the negative choice of square root:

$\gamma = -R - (1 + R^2)^{\frac{1}{2}} < -R - 1$ \hfill (3-28)

Hence $|\gamma| > 1$ for any choice of p and r. Thus eq. (3-27) is unconditionally unstable, and we must reject trying it, in spite of its attractive error estimate and ease of computation.

The Dufort-Frankel approximation

This is another three-level difference equation, similar to (3-27), but the central term $U_{i,n}$ in the second difference is replaced by the time average $(U_{i,n+1} + U_{i,n-1})/2$:

$$\frac{U_{i+1,n} - (U_{i,n+1} + U_{i,n-1}) + U_{i-1,n}}{\Delta x^2} = \frac{U_{i,n+1} - U_{i,n-1}}{2\Delta t} \qquad (3\text{-}29)$$

Let us first consider its stability. By substituting eq. (3-8), we obtain:

$$2(\Delta t/\Delta x^2)[\gamma^n \{\exp(ip\Delta x) + \exp(-ip\Delta x)\} - \{\gamma^{n+1} + \gamma^{n-1}\}] = \gamma^{n+1} - \gamma^{n-1}$$

Use of eq. (3-10) gives:

$$2(\Delta t/\Delta x^2)[\gamma^n(2\cos p\Delta x) - (\gamma^{n+1} + \gamma^{n-1})] = \gamma^{n+1} - \gamma^{n-1}$$

Divide by γ^{n-1} and rearrange to obtain:

$$\gamma^2(1 + 2r) - \gamma r(4\cos p\Delta x) + (2r - 1) = 0 \qquad (3\text{-}30)$$

$$\gamma = \frac{2r\cos\theta \pm (1 - 4r^2\sin^2\theta)^{1/2}}{1 + 2r} \qquad (3\text{-}31)$$

where $r = \Delta t/\Delta x^2$ and $\theta = p\Delta x$.

We now consider the two possibilities:
(1) $4r^2\sin^2\theta \geqslant 1$. Using $|a + ib|^2 = a^2 + b^2$, we can show that $|\gamma^2| = (2r - 1)/(2r + 1)$ and that $|\gamma|^2 < 1$.
(2) $4r^2\sin^2\theta \leqslant 1$. Using $|1 - 4r^2\sin^2\theta| < 1$ and the inequality $|a + b| \leqslant |a| + |b|$, we find that:

$$|\gamma| \leqslant \frac{|2r\cos\theta| + 1}{2r + 1} \leqslant 1$$

Hence eq. (3-29) is unconditionally stable.

However, a truncation error analysis shows the Dufort-Frankel approximation to be somewhat unusual. Following eq. (2-23), we write

$$\epsilon_L = \frac{u_{i+1,n} - (u_{i,n+1} + u_{i,n-1}) + u_{i-1,n}}{\Delta x^2} - \frac{u_{i,n+1} - u_{i,n-1}}{2\Delta t} - \frac{\partial^2 u}{\partial x^2} + \frac{\partial u}{\partial t}$$

In addition to using (2-10), i.e.:

$$\frac{\partial^2 u}{\partial x^2} = \frac{u_{i+1,n} - 2u_{in} + u_{i-1,n}}{\Delta x^2} - \frac{\Delta x^2}{12}\frac{\partial^4 u}{\partial x^4}$$

and rewriting (2-8) as:

$$\frac{\partial u}{\partial t} = \frac{u_{i,n+1} - u_{i,n-1}}{2\Delta t} - \frac{\Delta t^2}{6}\frac{\partial^3 u}{\partial t^3}$$

we also add the following two Taylor expansions together:

$$u_{i,n+1} = u_{in} + \Delta t\frac{\partial u}{\partial t} + \frac{\Delta t^2}{2}\frac{\partial^2 u}{\partial t^2}(t')$$

$$u_{i,n-1} = u_{in} - \Delta t \frac{\partial u}{\partial t} + \frac{\Delta t^2}{2} \frac{\partial^2 u}{\partial t^2}(t'')$$

to obtain:

$$u_{i,n+1} + u_{i,n-1} = 2u_{in} + \Delta t^2 \frac{\partial^2 u}{\partial t^2}(t''')$$

or:

$$\frac{u_{i,n+1} + u_{i,n-1}}{\Delta x^2} = \frac{2u_{in}}{\Delta x^2} + \frac{\Delta t^2}{\Delta x^2} \frac{\partial^2 u}{\partial t^2}$$

Then:

$$\epsilon_L = \frac{\Delta x^2}{12} \frac{\partial^4 u}{\partial x^4} - \frac{\Delta t^2}{6} \frac{\partial^3 u}{\partial t^3} - \frac{\Delta t^2}{\Delta x^2} \frac{\partial^2 u}{\partial t^2}$$

$$\epsilon_L = \mathcal{O}(\Delta x^2) + \mathcal{O}(\Delta t^2) + \mathcal{O}(\Delta t^2/\Delta x^2) \tag{3-32}$$

In order to get convergence similar to that of the forward- or backward-difference equations, it is necessary that Δt be kept proportional to Δx^2 (or smaller) as both Δt and Δx are reduced to zero. If Δt were kept proportional to Δx, we see that ϵ_L would remain constant (i.e., $\mathcal{O}(1)$), indicating lack of convergence. This indicates that Dufort-Frankel must be used with great care and that, probably, Δt should be kept relatively small, even for practically large values of Δx.

MULTIDIMENSIONAL PROBLEMS

Finite-difference methods for parabolic equations in several space variables can be separated into two categories: (1) extensions of the difference equations described above, and (2) alternating-direction methods, which have no analog in one space dimension. In this section, we will develop the extensions of the previous equations and confine our attention to two space variables, as that will be typical of the general multidimensional problem.

Forward-difference equation

Consider the differential equation:

$$\frac{\partial^2 u}{\partial x^2} + \frac{\partial^2 u}{\partial y^2} = \frac{\partial u}{\partial t} \tag{3-33}$$

The forward-difference equation (3-4) easily generalizes to:

$$U_{i,j,n+1} = U_{ijn} + \Delta t \Delta_x^2 U_{ijn} + \Delta t \Delta_y^2 U_{ijn} \tag{3-34}$$

where:

$$\Delta_x^2 U_{ijn} = (U_{i+1,j,n} - 2U_{ijn} + U_{i-1,j,n})/\Delta x^2$$

$$\Delta_y^2 U_{ijn} = (U_{i,j+1,n} - 2U_{ijn} + U_{i,j-1,n})/\Delta y^2 \tag{3-35}$$

To analyze the stability, we assume an error of the form:

$$\epsilon_{ijn} = \gamma^n \exp(ipi\Delta x) \exp(iqj\Delta y) \tag{3-36}$$

Substituting eq. (3-36) into (3-34) and cancelling common factors leads now to:

$$\gamma = 1 - 4(\Delta t/\Delta x^2) \sin^2(p\Delta x/2) - 4(\Delta t/\Delta y^2) \sin^2(q\Delta y/2) \tag{3-37}$$

To satisfy the von Neumann criterion for stability, we require:

$$-1 \leqslant \gamma \leqslant 1$$

The right-hand inequality is satisfied trivially, while the left-hand inequality leads to:

$$-2 \leqslant -4(\Delta t/\Delta x^2) \sin^2(p\Delta x/2) - (\Delta t/\Delta y^2) \sin^2(q\Delta y/2)$$

Maximizing the sines of $(p\Delta x/2)$ and $(q\Delta y/2)$ will require the smallest value of Δt in order to satisfy this inequality. As a conservative approximation, we can take:

$$\max \sin^2(p\Delta x/2) \approx \max \sin^2(q\Delta y/2) \approx 1$$

giving:

$$\Delta t \left(\frac{1}{\Delta x^2} + \frac{1}{\Delta y^2} \right) \leqslant \frac{1}{2} \tag{3-38}$$

For $\Delta x = \Delta y$, we have:

$$r = \Delta t/\Delta x^2 = \Delta t/\Delta y^2 \leqslant \tfrac{1}{4} \tag{3-39}$$

as the requirement for stability. By comparing with eq. (3-14), we see that the critical time step for a two-dimensional problem is only half that for a one-dimensional problem when the forward-difference equation is used.

Implicit difference equations

Let us consider use of two-dimensional implicit equations to achieve unconditional stability. The backward-difference equation would be:

$$\Delta_x^2 U_{i,j,n+1} + \Delta_y^2 U_{i,j,n+1} = \frac{U_{i,j,n+1} - U_{ijn}}{\Delta t} \tag{3-40}$$

Substituting eq. (3-36) and cancelling common factors leads to:

$$\gamma = \frac{1}{1 + 4(\Delta t/\Delta x^2) \sin^2(p\Delta x/2) + 4(\Delta t/\Delta y^2) \sin^2(q\Delta y/2)} \tag{3-41}$$

which is always less than one in magnitude; hence, eq. (3-40) is unconditionally stable.

The Crank-Nicolson difference analog of eq. (3-33) is also unconditionally stable. The difference equation is:

$$\tfrac{1}{2}[\Delta_x^2 U_{i,j,n+1} + \Delta_y^2 U_{i,j,n+1} + \Delta_x^2 U_{ijn} + \Delta_y^2 U_{ijn}] = \frac{U_{i,j,n+1} - U_{ijn}}{\Delta t} \qquad (3\text{-}42)$$

Substituting eq. (3-36) and cancelling common factors leads to:

$$\gamma = \frac{1 - [2(\Delta t/\Delta x^2)\sin^2(p\Delta x/2) + 2(\Delta t/\Delta y^2)\sin^2(q\Delta y/2)]}{1 + [2(\Delta t/\Delta x^2)\sin^2(p\Delta x/2) + 2(\Delta t/\Delta y^2)\sin^2(q\Delta y/2)]} \qquad (3\text{-}43)$$

which is always less than one in magnitude.

While the implicit difference equations (3-40) and (3-42) are unconditionally stable, they are much more difficult to solve than their one-dimensional counterparts. Their use gives rise to systems of simultaneous equations that are of the same form as those arising in connection with solutions of elliptic problems, which we will be studying in Chapter 5 in considerable detail. Suffice it to say here that direct solution of this system of equations involves much more computational labor than that associated with the simpler tridiagonal system of equations, and hence the solution is usually attempted by some method of iteration.

It will be helpful for us, at this point, to list the truncation errors associated with the forward, backward, and Crank-Nicolson difference equations, in order to have a basis for comparison with alternating-direction methods discussed in the following section. By using the method outlined in Chapter 2, we can find that the local truncation error of the two-dimensional forward-difference equation (3-34) is:

F-D: $\quad \epsilon_L = \mathcal{O}(\Delta x^2) + \mathcal{O}(\Delta y^2) + \mathcal{O}(\Delta t) \qquad (3\text{-}44)$

The local truncation error of the two-dimensional backward-difference equation (3-40) is the same:

B-D: $\quad \epsilon_L = \mathcal{O}(\Delta x^2) + \mathcal{O}(\Delta y^2) + \mathcal{O}(\Delta t) \qquad (3\text{-}45)$

while the local truncation error of the two-dimensional Crank-Nicolson difference equation (3-42) is:

C-N: $\quad \epsilon_L = \mathcal{O}(\Delta x^2) + \mathcal{O}(\Delta y^2) + \mathcal{O}(\Delta t^2) \qquad (3\text{-}46)$

ALTERNATING-DIRECTION METHODS

The Peaceman-Rachford method

As Douglas (1961, p. 37) points out, the use of implicit difference equations is motivated by several desires. The primary one is to obtain unconditional stability. Better accuracy is also sought. However, as a practical matter, it is highly desirable that the resulting simultaneous equations be easy to solve. While the implicit equations for a single space variable satisfy these wishes reasonably well, the implicit equations so far considered for two or more space variables yield systems of equations that are quite difficult to solve. The alternating-direction method was first proposed by Peaceman and Rachford (1955) for two space variables, and extended to three space variables by Douglas and Rachford (1956), by Brian (1961), and by Douglas (1962), to simplify the solution problem while preserving unconditional stability and reasonable accuracy.

Consider first the heat equation (3-33) on the unit square, with $\Delta x = \Delta y$. The solution of tridiagonal linear equations is quite simple, as we have found. If the simultaneous equations are to be of this form, then *only one* of the spatial difference quotients can be evaluated at t_{n+1}. This leads to the difference equation:

$$\Delta_x^2 U_{i,j,n+1} + \Delta_y^2 U_{ijn} = \frac{U_{i,j,n+1} - U_{ijn}}{\Delta t} \tag{3-47}$$

Let us consider the von Neumann stability of eq. (3-47). Assume an error of the form (3-36). Now:

$$\Delta_x^2 \epsilon_{ijn} = [\gamma^n \exp\{ip(i+1)\Delta x\} - 2\gamma^n \exp\{ipi\Delta x\} + \gamma^n \exp\{ip(i-1)\Delta x\}]$$
$$\cdot \exp\{iqj\Delta y\}/\Delta x^2$$
$$= \gamma^n \exp\{ipi\Delta x\} \exp\{iqj\Delta y\}[\exp(ip\Delta x) - 2 + \exp(-ip\Delta x)]/\Delta x^2$$
$$= -(4/\Delta x^2)\epsilon_{ijn} \sin^2(p\Delta x/2) \tag{3-48a}$$

Similarly:

$$\Delta_y^2 \epsilon_{ijn} = -(4/\Delta y^2)\epsilon_{ijn} \sin^2(q\Delta y/2) \tag{3-48b}$$

Therefore, substitution of eqs. (3-36) and (3-48) into (3-47) gives:

$$-\gamma^{n+1}(4/\Delta x^2) \sin^2(p\Delta x/2) - \gamma^n(4/\Delta y^2) \sin^2(q\Delta y/2) = (\gamma^{n+1} - \gamma^n)/\Delta t$$

Let $r = \Delta t/\Delta x^2 = \Delta t/\Delta y^2$. Then:

$$\gamma^{n+1}[1 + 4r \sin^2(p\Delta x/2)] = \gamma^n[1 - 4r \sin^2(q\Delta y/2)]$$

$$\frac{\gamma^{n+1}}{\gamma^n} = \frac{1 - 4r \sin^2(q\Delta y/2)}{1 + 4r \sin^2(p\Delta x/2)} \tag{3-49}$$

If $r > \frac{1}{2}$, $p = 1$, and $q = J - 1$, then $|\gamma| > 1$, so that eq. (3-47) is unstable, at least for $r > \frac{1}{2}$. Suppose we had evaluated $\Delta_y^2 U$ instead of $\Delta_x^2 U$ at t_{n+1}. Then the difference equation would be:

$$\Delta_x^2 U_{ijn} + \Delta_y^2 U_{i,j,n+1} = \frac{U_{i,j,n+1} - U_{ijn}}{\Delta t} \qquad (3\text{-}50)$$

Instead of eq. (3-49), the stability ratio would be:

$$\frac{\gamma^{n+1}}{\gamma^n} = \frac{1 - 4r \sin^2 (p\Delta x/2)}{1 + 4r \sin^2 (q\Delta y/2)} \qquad (3\text{-}51)$$

which again yields $|\gamma| > 1$ for some r, p, and q, so that eq. (3-50) is also unstable for $r > \frac{1}{2}$.

Consider taking one time step using eq. (3-47) and then one time step using (3-50). Equation (3-49) stays the same, but (3-51) becomes:

$$\frac{\gamma^{n+2}}{\gamma^{n+1}} = \frac{1 - 4r \sin^2 (p\Delta x/2)}{1 + 4r \sin^2 (q\Delta y/2)} \qquad (3\text{-}52)$$

Multiplying eqs. (3-49) and (3-52) gives the stability ratio for the double step:

$$\frac{\gamma^{n+2}}{\gamma^n} = \frac{1 - 4r \sin^2 (p\Delta x/2)}{1 + 4r \sin^2 (p\Delta x/2)} \cdot \frac{1 - 4r \sin^2 (q\Delta y/2)}{1 + 4r \sin^2 (q\Delta y/2)} \qquad (3\text{-}53)$$

which is less than one in magnitude for any size time step for any p and q. The *alternate* use of eqs. (3-47) and (3-50) results in a stable procedure, as was first shown by Peaceman and Rachford (1955).

Since we are interested in the solution after the double time step, let us alter Δt to correspond to the double step, and use $n + \frac{1}{2}$ to indicate the intermediate values:

$$\Delta_x^2 U_{i,j,n+\frac{1}{2}} + \Delta_y^2 U_{ijn} = \frac{U_{i,j,n+\frac{1}{2}} - U_{ijn}}{\Delta t/2}$$

$$\Delta_x^2 U_{i,j,n+\frac{1}{2}} + \Delta_y^2 U_{i,j,n+1} = \frac{U_{i,j,n+1} - U_{i,j,n+\frac{1}{2}}}{\Delta t/2} \qquad (3\text{-}54)$$

To facilitate studying the accuracy of eq. (3-54), let us eliminate $U_{i,j,n+\frac{1}{2}}$ to obtain what we might call an "overall" equation. First we add and subtract the two equations (3-54):

$$2\Delta_x^2 U_{i,j,n+\frac{1}{2}} + \Delta_y^2 (U_{ijn} + U_{i,j,n+1}) = 2(U_{i,j,n+1} - U_{ijn})/\Delta t \qquad (3\text{-}55)$$

$$\Delta_y^2 (U_{i,j,n+1} - U_{ijn}) = 2(U_{i,j,n} + U_{i,j,n+1} - 2U_{i,j,n+\frac{1}{2}})/\Delta t \qquad (3\text{-}56)$$

Solve eq. (3-56) for $U_{i,j,n+\frac{1}{2}}$, take its second x-difference and substitute into (3-55):

$$U_{i,j,n+\frac{1}{2}} = \tfrac{1}{2}(U_{ijn} + U_{i,j,n+1}) - (\Delta t/4)\Delta_y^2(U_{i,j,n+1} - U_{ijn})$$

$$\Delta_x^2 U_{i,j,n+\frac{1}{2}} = \tfrac{1}{2}\Delta_x^2(U_{ijn} + U_{i,j,n+1}) - (\Delta t/4)\Delta_x^2\Delta_y^2(U_{i,j,n+1} - U_{ijn})$$

$$\Delta_x^2(U_{ijn} + U_{i,j,n+1}) - (\Delta t/2)\Delta_x^2\Delta_y^2(U_{i,j,n+1} - U_{ijn}) +$$

$$+ \Delta_y^2(U_{ijn} + U_{i,j,n+1}) = 2(U_{i,j,n+1} - U_{ijn})/\Delta t$$

$$\tfrac{1}{2}(\Delta_x^2 + \Delta_y^2)(U_{ijn} + U_{i,j,n+1}) = \frac{U_{i,j,n+1} - U_{ijn}}{\Delta t}$$

$$+ (\Delta t/4)\Delta_x^2\Delta_y^2(U_{i,j,n+1} - U_{ijn}) \tag{3-57}$$

As Douglas (1961, p. 38) noted, eq. (3-57) is a perturbation of the Crank-Nicolson equation (3-42). Like it, (3-57) is also second-order correct both in space and time; that is, the error is $\mathcal{O}[(\Delta x)^2 + (\Delta y)^2 + (\Delta t)^2]$.

When we try to extend the Peaceman-Rachford method to three dimensions, we get the following difference equations. To simplify the writing, let us suppress i and j in the index:

$$\Delta_x^2 U_{n+\tfrac{1}{3}} + \Delta_y^2 U_n + \Delta_z^2 U_n = \frac{U_{n+\tfrac{1}{3}} - U_n}{\Delta t/3}$$

$$\Delta_x^2 U_{n+\tfrac{1}{3}} + \Delta_y^2 U_{n+\tfrac{2}{3}} + \Delta_z^2 U_n = \frac{U_{n+\tfrac{2}{3}} - U_{n+\tfrac{1}{3}}}{\Delta t/3}$$

$$\Delta_x^2 U_{n+\tfrac{1}{3}} + \Delta_y^2 U_{n+\tfrac{2}{3}} + \Delta_z^2 U_{n+1} = \frac{U_{n+1} - U_{n+\tfrac{2}{3}}}{\Delta t/3}$$

A stability analysis, after much messy manipulation, yields:

$$\frac{\gamma^{n+1}}{\gamma^n} = 1 + \frac{3(Y-9)(X+Y+Z)}{(3+X)(3+Y)(3+Z)}$$

where:

$X = 4(\Delta t/\Delta x^2) \sin^2(p\Delta x/2)$
$Y = 4(\Delta t/\Delta y^2) \sin^2(q\Delta y/2)$
$Z = 4(\Delta t/\Delta z^2) \sin^2(s\Delta z/2)$

If $X > 6$ and $Y = Z = 0$, then $\gamma < -1$. Thus $\Delta t/\Delta x^2 > \tfrac{3}{2}$ implies instability, so that the three-dimensional version of the Peaceman-Rachford method is not unconditionally stable.

The Douglas-Rachford method

The first unconditionally stable alternating-direction method for three dimensions was proposed by Douglas and Rachford (1956):

$$\Delta_x^2 U_{n+\tfrac{1}{3}} + \Delta_y^2 U_n + \Delta_z^2 U_n = \frac{U_{n+\tfrac{1}{3}} - U_n}{\Delta t} \tag{3-58a}$$

$$\Delta_y^2 U_{n+\frac{2}{3}} = \Delta_y^2 U_n + \frac{U_{n+\frac{2}{3}} - U_{n+\frac{1}{3}}}{\Delta t} \tag{3-58b}$$

$$\Delta_z^2 U_{n+1} = \Delta_z^2 U_n + \frac{U_{n+1} - U_{n+\frac{2}{3}}}{\Delta t} \tag{3-58c}$$

The second and third equations (3-58) might be regarded as "corrections" to the first equation. A standard stability analysis shows eq. (3-58) to be unconditionally stable; it yields the result:

$$\gamma = \frac{1 + XY + XZ + YZ + XYZ}{1 + XY + XZ + YZ + XYZ + X + Y + Z} < 1$$

We eliminate $U_{n+\frac{1}{3}}$ and $U_{n+\frac{2}{3}}$ to obtain an overall equation as follows. First solve eq. (3-58c) for $U_{n+\frac{2}{3}}$ and difference with respect to y.

$$U_{n+\frac{2}{3}} = U_{n+1} - \Delta t \Delta_z^2 (U_{n+1} - U_n)$$
$$\Delta_y^2 U_{n+\frac{2}{3}} = \Delta_y^2 U_{n+1} - \Delta t \Delta_y^2 \Delta_z^2 (U_{n+1} - U_n)$$

Solve eq. (3-58b) for $U_{n+\frac{1}{3}}$, substitute both of the above equations, and difference with respect to x:

$$U_{n+\frac{1}{3}} = U_{n+\frac{2}{3}} - \Delta t \Delta_y^2 U_{n+\frac{2}{3}} + \Delta t \Delta_y^2 U_n$$
$$U_{n+\frac{1}{3}} = U_{n+1} - \Delta t \Delta_z^2 (U_{n+1} - U_n) - \Delta t [\Delta_y^2 U_{n+1} - \Delta t \Delta_y^2 \Delta_z^2 (U_{n+1} - U_n)]$$
$$+ \Delta t \Delta_y^2 U_n$$
$$\Delta_x^2 U_{n+\frac{1}{3}} = \Delta_x^2 U_{n+1} - \Delta t \Delta_x^2 \Delta_z^2 (U_{n+1} - U_n) - \Delta t \Delta_x^2 \Delta_y^2 U_{n+1}$$
$$+ \Delta t^2 \Delta_x^2 \Delta_y^2 \Delta_z^2 (U_{n+1} - U_n) + \Delta t \Delta_x^2 \Delta_y^2 U_n$$

Substitute the above two equations into eq. (3-58a)

$$\Delta_x^2 U_{n+1} - \Delta t \Delta_x^2 \Delta_z^2 (U_{n+1} - U_n) - \Delta t \Delta_x^2 \Delta_y^2 U_{n+1} + \Delta t^2 \Delta_x^2 \Delta_y^2 \Delta_z^2 (U_{n+1} - U_n)$$
$$+ \Delta t \Delta_x^2 \Delta_y^2 U_n + \Delta_y^2 U_n + \Delta_z^2 U_n = (U_{n+1}/\Delta t) - \Delta_z^2 (U_{n+1} - U_n)$$
$$- \Delta_y^2 U_{n+1} + \Delta t \Delta_y^2 \Delta_z^2 (U_{n+1} - U_n) + \Delta_y^2 U_n - (U_n/\Delta t)$$

$$\Delta_x^2 U_{n+1} + \Delta_y^2 U_{n+1} + \Delta_z^2 U_{n+1} = \frac{U_{n+1} - U_n}{\Delta t} - \Delta t^2 \Delta_x^2 \Delta_y^2 \Delta_z^2 (U_{n+1} - U_n)$$
$$+ \Delta t (\Delta_x^2 \Delta_y^2 + \Delta_x^2 \Delta_z^2 + \Delta_y^2 \Delta_z^2)(U_{n+1} - U_n) \tag{3-59}$$

We see that eq. (3-59) is a perturbation of the backward-difference equation (3-40). Like the backward-difference equation, it has an error term of $\mathcal{O}[(\Delta x)^2 + (\Delta y)^2 + (\Delta z)^2 + \Delta t]$.

An interesting aspect of the two-dimensional version of the Douglas-Rachford method is that if it is used for half a time step (i.e., for $\Delta t/2$), and then "extrapolated" to a full time step, the result is identical with the Peaceman-Rachford procedure for two dimensions. The two-dimensional version of Douglas-Rachford for half a time step is:

$$\Delta_x^2 U_{n+\frac{1}{4}} + \Delta_y^2 U_n = \frac{U_{n+\frac{1}{4}} - U_n}{\Delta t/2}$$

$$\Delta_y^2 U_{n+\frac{1}{2}} = \Delta_y^2 U_n + \frac{U_{n+\frac{1}{2}} - U_{n+\frac{1}{4}}}{\Delta t/2} \tag{3-60}$$

The extrapolation equation is:

$$U_{n+1} = U_n + 2(U_{n+\frac{1}{2}} - U_n) \tag{3-61}$$

Now the overall equation for (3-60) can be obtained from (3-59) by setting $\Delta_z^2 = 0$ and using $\Delta t/2$ for Δt:

$$(\Delta_x^2 + \Delta_y^2)U_{n+\frac{1}{2}} = \frac{U_{n+\frac{1}{2}} - U_n}{\Delta t/2} + (\Delta t/2)\Delta_x^2 \Delta_y^2 (U_{n+\frac{1}{2}} - U_n) \tag{3-62}$$

From eq. (3-61), we have $(U_{n+\frac{1}{2}} - U_n) = (U_{n+1} - U_n)/2$ and $U_{n+\frac{1}{2}} = (U_n + U_{n+1})/2$. Substituting into the above gives (3-57), the overall equation for the Peaceman-Rachford method.

The Brian and Douglas methods

Brian (1961) used the same extrapolation approach to the three-dimensional Douglas-Rachford method to obtain a more accurate three-dimensional method. Thus the difference equation is:

$$\Delta_x^2 U_{n+\frac{1}{6}} + \Delta_y^2 U_n + \Delta_z^2 U_n = \frac{U_{n+\frac{1}{6}} - U_n}{\Delta t/2}$$

$$\Delta_y^2 U_{n+\frac{2}{6}} = \Delta_y^2 U_n + \frac{U_{n+\frac{2}{6}} - U_{n+\frac{1}{6}}}{\Delta t/2}$$

$$\Delta_z^2 U_{n+\frac{1}{2}} = \Delta_z^2 U_n + \frac{U_{n+\frac{1}{2}} - U_{n+\frac{2}{6}}}{\Delta t/2} \tag{3-63}$$

Equation (3-61) still serves as the extrapolation equation. Making the appropriate changes in the overall equation (3-59) gives:

$$(\Delta_x^2 + \Delta_y^2 + \Delta_z^2)U_{n+\frac{1}{2}} = \frac{U_{n+\frac{1}{2}} - U_n}{\Delta t/2} - (\Delta t^2/4)\Delta_x^2 \Delta_y^2 \Delta_z^2 (U_{n+\frac{1}{2}} - U_n)$$

$$+ (\Delta t/2)(\Delta_x^2 \Delta_y^2 + \Delta_x^2 \Delta_z^2 + \Delta_y^2 \Delta_z^2)(U_{n+\frac{1}{2}} - U_n)$$

Again using $(U_{n+\frac{1}{2}} - U_n) = (U_{n+1} - U_n)/2$ and $U_{n+\frac{1}{2}} = (U_n + U_{n+1})/2$, we have:

$$\tfrac{1}{2}(\Delta_x^2 + \Delta_y^2 + \Delta_z^2)(U_n + U_{n+1}) = \frac{U_{n+1} - U_n}{\Delta t} - (\Delta t^2/8)\Delta_x^2 \Delta_y^2 \Delta_z^2 (U_{n+1} - U_n)$$

$$+ (\Delta t/4)(\Delta_x^2 \Delta_y^2 + \Delta_x^2 \Delta_z^2 + \Delta_y^2 \Delta_z^2)(U_{n+1} - U_n) \tag{3-64}$$

Douglas (1962) approached the problem of obtaining a more accurate three-dimensional procedure by setting up an alternating-direction modification of the Crank-Nicolson method from the beginning. (The two-dimensional version of the following treatment turns out to be equivalent to the Peaceman-Rachford method.) First evaluate the x-difference at $t_{n+\frac{1}{2}}$ to obtain $U_{n+\frac{1}{3}}$, then move the evaluation of the y-difference to $t_{n+\frac{1}{2}}$ when obtaining $U_{n+\frac{2}{3}}$; finally, move the evaluation of the z-difference to $t_{n+\frac{1}{2}}$ when obtaining U_{n+1}.

$$\tfrac{1}{2}\Delta_x^2(U_{n+\frac{1}{3}} + U_n) + \Delta_y^2 U_n + \Delta_z^2 U_n = \frac{U_{n+\frac{1}{3}} - U_n}{\Delta t}$$

$$\tfrac{1}{2}\Delta_x^2(U_{n+\frac{1}{3}} + U_n) + \tfrac{1}{2}\Delta_y^2(U_{n+\frac{2}{3}} + U_n) + \Delta_z^2 U_n = \frac{U_{n+\frac{2}{3}} - U_n}{\Delta t}$$

$$\tfrac{1}{2}\Delta_x^2(U_{n+\frac{1}{3}} + U_n) + \tfrac{1}{2}\Delta_y^2(U_{n+\frac{2}{3}} + U_n) + \tfrac{1}{2}\Delta_z^2(U_{n+1} + U_n)$$
$$= \frac{U_{n+1} - U_n}{\Delta t} \tag{3-65}$$

A more convenient form for calculation may be obtained by rearranging the first equation and by subtracting the first two and last two equations of (3-65):

$$\tfrac{1}{2}\Delta_x^2(U_{n+\frac{1}{3}} - U_n) + (\Delta_x^2 + \Delta_y^2 + \Delta_z^2)U_n = \frac{U_{n+\frac{1}{3}} - U_n}{\Delta t}$$

$$\tfrac{1}{2}\Delta_y^2(U_{n+\frac{2}{3}} - U_n) = \frac{U_{n+\frac{2}{3}} - U_{n+\frac{1}{3}}}{\Delta t}$$

$$\tfrac{1}{2}\Delta_z^2(U_{n+1} - U_n) = \frac{U_{n+1} - U_{n+\frac{2}{3}}}{\Delta t} \tag{3-66}$$

Elimination of $U_{n+\frac{1}{3}}$ and $U_{n+\frac{2}{3}}$ from eq. (3-66) leads to the same overall equation as (3-64), showing that Brian's method and Douglas's method are equivalent (although the intermediate numbers are different for the two methods). Equation (3-64) is a perturbation of the Crank-Nicolson equation for three dimensions, and also has an error of $\mathcal{O}[(\Delta x)^2 + (\Delta y)^2 + (\Delta z)^2 + (\Delta t)^2]$. Thus the Brian-Douglas procedures possess the desired accuracy, as compared with the Douglas-Rachford method, which is only $\mathcal{O}[(\Delta x)^2 + (\Delta y)^2 + (\Delta z)^2 + \Delta t]$.

The *analyses* of alternating-direction procedures cannot be extended to nonrectangular regions or to cases where the coefficients of the differential equations are not constant. This does not mean that alternating-direction methods themselves fail for such problems; indeed, the methods have been found very useful for the more complex problems.

NOMENCLATURE

Symbols

a_i = off-diagonal coefficient of tridiagonal system of equations
b_i = diagonal coefficient of tridiagonal system of equations
c_i = off-diagonal coefficient of tridiagonal system of equations
d_i = right-hand side in tridiagonal system of equations
f_0 = function specifying initial condition
g_1 = function specifying boundary condition at $x = x_1$
g_2 = function specifying boundary condition at $x = x_2$
$i = (-1)^{\frac{1}{2}}$
I = number of grid points in x-direction
p = index in Fourier expansion of error
q = index in Fourier expansion of error
$r = \Delta t/\Delta x^2$ or $\Delta t/\Delta y^2$
s = index in Fourier expansion of error
t = time
u = general function
U = difference solution for function u
w_i = intermediate variable in solution of tridiagonal system of equations
x = distance
$X = 4(\Delta t/\Delta x^2) \sin^2 (p\Delta x/2)$
y = distance
$Y = 4(\Delta t/\Delta y^2) \sin^2 (q\Delta y/2)$
z = distance
$Z = 4(\Delta t/\Delta z^2) \sin^2 (s\Delta z/2)$
β_i = intermediate variable in solution of tridiagonal system of equations
γ_i = intermediate variable in solution of tridiagonal system of equations
$\gamma = \gamma_p$ = amplification factor for stability analysis
$\gamma^n = \gamma_p^n$ = coefficient in Fourier expansion of error
Δt = increment of t
Δx = increment of x
Δy = increment of y
Δz = increment of z
$\Delta_x^2 U_{ij}$ = second-difference quotient in x-direction, equals $(U_{i+1,j} - 2U_{ij} + U_{i-1,j})/\Delta x^2$
$\Delta_y^2 U_{ij}$ = second-difference quotient in y-direction, equals $(U_{i,j+1} - 2U_{ij} + U_{i,j-1})/\Delta y^2$
$\Delta_z^2 U_{ijk}$ = second-difference quotient in z-direction, equals $(U_{i,j,k+1} - 2U_{ijk} + U_{i,j,k-1})/\Delta z^2$
ϵ_L = local truncation error
$\theta = p\Delta x$

Subscripts

i index in x-direction
j index in y-direction
k index in z-direction
n index in t-direction (i.e., time)
p index in Fourier expansion

CHAPTER 4

NUMERICAL SOLUTION OF FIRST-ORDER HYPERBOLIC PROBLEMS IN ONE DEPENDENT VARIABLE

INTRODUCTION

The first-order hyperbolic problem we shall consider is the following *nonlinear* convection equation in one space dimension:

$$-v \frac{\partial}{\partial x} f(u) = \frac{\partial u}{\partial t} \tag{4-1}$$

where f is some function of the dependent variable, u. We observe that eq. (4-1) is a simplified version of eq. (1-63) (which was derived for two-phase flow), the simplifications being the assumptions of one-dimensional, *horizontal* flow and *zero* capillary pressure. The velocity, v, is assumed to be positive, corresponding to flow in the direction of increasing x.

In this chapter, we shall consider several simple difference equations for approximating (4-1), and examine them for stability and truncation error. An understanding of the behavior of these simple difference equations should lead to an improved understanding of some of the problems associated with the numerical solution of multiphase flow problems, the subject covered in Chapter 6.

DIFFERENCE EQUATIONS

Distance-weighting

The first step in obtaining a difference analog for eq. (4-1) is to approximate the space derivative, $\partial f/\partial x$, by:

$$\frac{\partial f}{\partial x} \approx \frac{F_{i+\frac{1}{2}} - F_{i-\frac{1}{2}}}{\Delta x} \tag{4-2}$$

Several options are available for choosing $F_{i+\frac{1}{2}}$. The simplest option is arithmetic averaging:

$$F_{i+\frac{1}{2}} = (F_i + F_{i+1})/2 = [f(U_i) + f(U_{i+1})]/2 \tag{4-3}$$

Another option, more commonly used, is:

$$F_{i+\frac{1}{2}} = F_i = f(U_i) \tag{4-4}$$

Since the point, x_i, is the *upstream* end of the interval between x_i and x_{i+1}, we refer to this as *upstream weighting* of the function f, while eq. (4-3) is referred to as *mid-point weighting*.

A weighting parameter, W, can be used to characterize the options:

upstream weighting: $W = 1$

mid-point weighting: $W = \frac{1}{2}$

downstream weighting: $W = 0$

Then eqs. (4-3) and (4-4) can be combined into the general equation:

$$F_{i+\frac{1}{2}} = (W)F_i + (1-W)F_{i+1} \tag{4-5}$$

Similarly:

$$F_{i-\frac{1}{2}} = (W)F_{i-1} + (1-W)F_i \tag{4-6}$$

Time-weighting

We can use a similar parametric approach to differentiating among explicit, implicit, and centered-in-time (e.g., Crank-Nicolson for the heat equation) difference equations, wherein the spatial derivatives are evaluated, respectively, at t_n, t_{n+1}, or $t_{n+\frac{1}{2}}$. If we use θ as the "time-weighting" parameter, with the values:

implicit: $\theta = 1$

centered-in-time: $\theta = \frac{1}{2}$

explicit: $\theta = 0$

then eq. (4-1) can be replaced by the difference analog:

$$-\frac{v}{\Delta x}[\theta(F_{i+\frac{1}{2},n+1} - F_{i-\frac{1}{2},n+1}) + (1-\theta)(F_{i+\frac{1}{2},n} - F_{i-\frac{1}{2},n})] = \frac{U_{i,n+1} - U_{i,n}}{\Delta t} \tag{4-7}$$

General form of difference equation

Substitution of eqs. (4-5) and (4-6) into eq. (4-7) gives the final, general, difference analog for eq. (4-1):

$$-(v\Delta t/\Delta x)[\theta\{(1-W)F_{i+1,n+1} + (2W-1)F_{i,n+1} - (W)F_{i-1,n+1}\}$$
$$+ (1-\theta)\{(1-W)F_{i+1,n} + (2W-1)F_{i,n} - (W)F_{i-1,n}\}]$$
$$= U_{i,n+1} - U_{i,n} \tag{4-8}$$

Nine particular cases of special interest are listed in Table IV, together

TABLE IV

Special cases of first-order difference equation

	Backward-in-distance $W = 1$	Centered-in-distance $W = \frac{1}{2}$	Forward-in-distance $W = 0$
Backward-in-time $\theta = 1$	always stable $\mathcal{D}_{num} = \frac{1}{2}vf'\Delta x(\lambda + 1)$	always stable $\mathcal{D}_{num} = \frac{1}{2}vf'\Delta x(\lambda)$	stable if $\lambda \geq 1$ $\mathcal{D}_{num} = \frac{1}{2}vf'\Delta x(\lambda - 1)$
Centered-in-time $\theta = \frac{1}{2}$	always stable $\mathcal{D}_{num} = \frac{1}{2}vf'\Delta x$	neutrally stable $\mathcal{D}_{num} = 0$	always unstable $\mathcal{D}_{num} = -\frac{1}{2}vf'\Delta x$
Forward-in-time $\theta = 0$	stable if $\lambda \leq 1$ $\mathcal{D}_{num} = \frac{1}{2}vf'\Delta x(1 - \lambda)$	always unstable $\mathcal{D}_{num} = -\frac{1}{2}vf'\Delta x(\lambda)$	always unstable $\mathcal{D}_{num} = -\frac{1}{2}vf'\Delta x(\lambda + 1)$

with schematic diagrams and descriptive names. Also included are conclusions regarding stability (which are derived in the next section) and the value of a variable, $\mathfrak{D}_{num}$, which is discussed in a subsequent section.

Linearization of difference equation

For any $\theta > 0$, eq. (4-8) is nonlinear, because of the dependence of $F_{i,n+1}$ on $U_{i,n+1}$. At this point, we shall not concern ourselves with the procedure for solving the nonlinear set of equations arising at each time step.

The nonlinearity can, however, make difficult the task of analyzing for stability and truncation error. To simplify this task, we shall assume (on a heuristic basis) that f is a straight-line function of u and, hence, that df/du is a constant. From this assumption we can write:

$$F_{i+1} - F_i = f'(U_{i+1} - U_i) \tag{4-9}$$

where $f' = df/du$. Letting

$$\lambda = vf'\Delta t/\Delta x \tag{4-10}$$

Equation (4-8) can then be written in the *linear* form:

$$-\lambda[\theta\{(1-W)U_{i+1,n+1} + (2W-1)U_{i,n+1} - (W)U_{i-1,n+1}\}$$
$$+ (1-\theta)\{(1-W)U_{i+1,n} + (2W-1)U_{i,n} - (W)U_{i-1,n}\}]$$
$$= U_{i,n+1} - U_{i,n} \tag{4-11}$$

STABILITY

We proceed with a von Neumann stability analysis of eq. (4-11) in a manner similar to that developed in Chapter 3. By substituting eq. (3-8) into (4-11) and cancelling the common factor $\exp(ipi\Delta x)$, we obtain:

$$-\lambda[\theta\gamma^{n+1}\{(1-W)\exp(ip\Delta x) + (2W-1) - (W)\exp(-ip\Delta x)\}$$
$$+ (1-\theta)\gamma^n\{(1-W)\exp(ip\Delta x) + (2W-1) - (W)\exp(-ip\Delta x)\}]$$
$$= \gamma^{n+1} - \gamma^n$$

We make use of the equation:

$$\exp(\pm ip\Delta x) = \cos(p\Delta x) \pm i\sin(p\Delta x)$$

to obtain:

$$-\lambda[\theta\gamma^{n+1} + (1-\theta)\gamma^n][(2W-1)\{1-\cos(p\Delta x)\} + i\sin(p\Delta x)]$$
$$= \gamma^{n+1} - \gamma^n$$

or:

$$\frac{\gamma^{n+1}}{\gamma^n} = \frac{1 - \lambda(1-\theta)(2W-1)\{1 - \cos(p\Delta x)\} - i\lambda(1-\theta)\sin(p\Delta x)}{1 + \lambda\theta(2W-1)\{1 - \cos(p\Delta x)\} + i\lambda\theta\sin(p\Delta x)} \quad (4\text{-}12)$$

That the amplification factor, $\gamma = \gamma^{n+1}/\gamma^n$, is a complex number is typical in the stability analysis of a first-order hyperbolic difference equation. The von Neumann condition for stability is that the modulus of the amplification factor be less than or equal to one. To obtain this modulus, we need merely to make use of the fact that if:

$$\gamma = \frac{a + ib}{c + id}$$

then:

$$|\gamma|^2 = \frac{|a + ib|^2}{|c + id|^2} = \frac{a^2 + b^2}{c^2 + d^2}$$

From eq. (4-12), then, we obtain:

$$|\gamma|^2 = \frac{[1 - \lambda(1-\theta)(2W-1)\{1 - \cos(p\Delta x)\}]^2 + \lambda^2(1-\theta)^2 \sin^2(p\Delta x)}{[1 + \lambda\theta(2W-1)\{1 - \cos(p\Delta x)\}]^2 + \lambda^2\theta^2 \sin^2(p\Delta x)}$$

$$(4\text{-}13)$$

Further analysis will be simplified if we consider, separately, the three cases of $W = \frac{1}{2}$, $W = 1$, and $W = 0$.

Stability of centered-in-distance equations

For $W = \frac{1}{2}$, eq. (4-13) simplifies to:

$$|\gamma|^2 = \frac{1 + \lambda^2(1-\theta)^2 \sin^2(p\Delta x)}{1 + \lambda^2\theta^2 \sin^2(p\Delta x)} \quad (4\text{-}14)$$

The requirement that $|\gamma|^2 \leq 1$ leads to:

$$1 + \lambda^2(1-\theta)^2 \sin^2(p\Delta x) \leq 1 + \lambda^2\theta^2 \sin^2(p\Delta x)$$

or:

$$(1-\theta)^2 \leq \theta^2$$

The inequality is satisfied for *any* $\theta > \frac{1}{2}$, and is *never* satisfied for $\theta < \frac{1}{2}$. Inspection of eq. (4-14) shows $|\gamma|$ is exactly one for $\theta = \frac{1}{2}$; we might refer to the centered-in-distance, centered-in-time difference equation, then, as being "neutrally stable". The centered-in-distance, backward-in-time equation ($\theta = 1$) is unconditionally stable, while the centered-in-distance, forward-in-time equation ($\theta = 0$) is unconditionally unstable.

Stability of backward-in-distance equations

For the case of $W = 1$, eq. (4-13) becomes:

$$|\gamma|^2 = \frac{[1 - \lambda(1 - \theta)\{1 - \cos(p\Delta x)\}]^2 + \lambda^2(1 - \theta)^2 \sin^2(p\Delta x)}{[1 + \lambda\theta\{1 - \cos(p\Delta x)\}]^2 + \lambda^2\theta^2 \sin^2(p\Delta x)} \quad (4\text{-}15)$$

It is immediately obvious that for $\theta = 1$ (backward-in-time), the numerator of eq. (4-15) is one, while the denominator is larger than one. Thus $|\gamma| < 1$, so this particular case is unconditionally stable.

For the more general case, we make use of the requirement that $|\gamma|^2 \leq 1$ to obtain the condition (for stability) that the numerator of eq. (4-15) be less than or equal to the denominator. Upon expansion and some cancellation, we obtain:

$$-2\lambda\{1 - \cos(p\Delta x)\} + \lambda^2(1 - 2\theta)\{1 - \cos(p\Delta x)\}^2$$
$$+ \lambda^2(1 - 2\theta)\sin^2(p\Delta x) \leq 0$$

which reduces to:

$$-2\lambda\{1 - \cos(p\Delta x)\}[1 - \lambda(1 - 2\theta)] \leq 0$$

Since $\lambda > 0$ and $\cos(p\Delta x) \leq 1$, stability then requires that:

$$1 - \lambda(1 - 2\theta) \geq 0 \quad (4\text{-}16)$$

This is automatically satisfied for any $\theta \geq \frac{1}{2}$.

When $\theta < \frac{1}{2}$, eq. (4-16) leads to a restriction on λ:

for stability: $\quad \lambda \leq \dfrac{1}{1 - 2\theta}$

unstable if: $\quad \lambda > \dfrac{1}{1 - 2\theta}$

In particular, when $\theta = 0$, corresponding to the important backward-in-distance, forward-in-time equation, the stability requirement is $\lambda \leq 1$, or:

$$\frac{vf'\Delta t}{\Delta x} \leq 1 \quad (4\text{-}17)$$

This restriction on Δt has a physical interpretation. In one time step, the fluid moves a distance equal to $v\Delta t$, and (ignoring the effect of f') eq. (4-17) requires that this distance be less than Δx.

Stability of forward-in-distance equations

For the case of $W = 0$, eq. (4-13) becomes:

$$|\gamma|^2 = \frac{[1 + \lambda(1-\theta)\{1 - \cos(p\Delta x)\}]^2 + \lambda^2(1-\theta)^2 \sin^2(p\Delta x)}{[1 - \lambda\theta\{1 - \cos(p\Delta x)\}]^2 + \lambda^2\theta^2 \sin^2(p\Delta x)} \qquad (4\text{-}18)$$

Again, the condition for stability is that the numerator be less than the denominator. After expansion and some cancellation, this leads to:

$$2\lambda\{1 - \cos(p\Delta x)\} + \lambda^2(1 - 2\theta)\{1 - \cos(p\Delta x)\}^2$$
$$+ \lambda^2(1 - 2\theta)\sin^2(p\Delta x) \leqslant 0$$

which reduces to:

$$2\lambda\{1 - \cos(p\Delta x)\}[1 + \lambda(1 - 2\theta)] \leqslant 0$$

Stability then requires that:

$$1 + \lambda(1 - 2\theta) \leqslant 0 \qquad (4\text{-}19)$$

This is *never* satisfied for any $\theta \leqslant \frac{1}{2}$.

When $\theta > \frac{1}{2}$, eq. (4-19) leads to the following restriction on λ (for stability):

$$\lambda \geqslant \frac{1}{2\theta - 1}$$

In particular, for $\theta = 1$ (corresponding to the forward-in-distance, backward-in-time equation), the stability requirement is $\lambda \geqslant 1$, or:

$$\frac{vf'\Delta t}{\Delta x} \geqslant 1 \qquad (4\text{-}20)$$

This is an unusual requirement for stability, since it requires that the time step be at least as large as some critical value. In practice, such a restriction would be extremely difficult to satisfy at all grid points, and could lead to excessively large time steps.

In summary, we can conclude that none of the forward-in-distance equations is useful, either because of their unconditional instability (when $\theta = 0$ or $\theta = \frac{1}{2}$), or because of their "inverse" conditional stability (when $\theta > \frac{1}{2}$).

TRUNCATION ERROR ANALYSIS – NUMERICAL DISPERSION

Local truncation error

To analyze the truncation error associated with the linear difference equation (4-11), we shall use the linearized form of the differential equation (4-1), which is:

$$-vf'\frac{\partial u}{\partial x} = \frac{\partial u}{\partial t} \tag{4-21}$$

We note that the expression:

$$(1-W)U_{i+1} + (2W-1)U_i - (W)U_{i-1}$$

which occurs in eq. (4-11) can also be written:

$$\tfrac{1}{2}(U_{i+1} - U_{i-1}) + (\tfrac{1}{2} - W)(U_{i+1} - 2U_i + U_{i-1})$$

Thus, by substituting eqs. (4-11) and (4-21) into the definition of local truncation error (eq. 2-23), we have:

$$\begin{aligned}\epsilon_L = -vf'\Bigg[\theta\bigg\{&\frac{u_{i+1,n+1} - u_{i-1,n+1}}{2\Delta x} \\ &+ \Delta x(\tfrac{1}{2} - W)\frac{u_{i+1,n+1} - 2u_{i,n+1} + u_{i-1,n+1}}{\Delta x^2}\bigg\} \\ + (1-\theta)\bigg\{&\frac{u_{i+1,n} - u_{i-1,n}}{2\Delta x} + \Delta x(\tfrac{1}{2} - W)\frac{u_{i+1,n} - 2u_{i,n} + u_{i-1,n}}{\Delta x^2}\bigg\}\Bigg] \\ -\frac{u_{i,n+1} - u_{i,n}}{\Delta t} &+ vf'\left(\frac{\partial u}{\partial x}\right)_{i,m} + \left(\frac{\partial u}{\partial t}\right)_{i,m}\end{aligned} \tag{4-22}$$

where we take:

$$m = (1-\theta)n + \theta(n+1)$$

We rewrite eq. (2-8)

$$\frac{u_{i+1} - u_{i-1}}{2\Delta x} = \left(\frac{\partial u}{\partial x}\right)_i + \mathcal{O}(\Delta x^2)$$

and eq. (2-10)

$$\frac{u_{i+1} - 2u_i + u_{i-1}}{\Delta x^2} = \left(\frac{\partial^2 u}{\partial x^2}\right)_i + \mathcal{O}(\Delta x^2)$$

to obtain:

$$\begin{aligned}\epsilon_L = -vf'\Bigg[&\theta\left(\frac{\partial u}{\partial x}\right)_{i,n+1} + \theta\Delta x(\tfrac{1}{2} - W)\left(\frac{\partial^2 u}{\partial x^2}\right)_{i,n+1} + (1-\theta)\left(\frac{\partial u}{\partial x}\right)_{i,n} \\ &+ (1-\theta)\Delta x(\tfrac{1}{2} - W)\left(\frac{\partial^2 u}{\partial x^2}\right)_{i,n}\Bigg] - \frac{u_{i,n+1} - u_{i,n}}{\Delta t} \\ &+ vf'\left(\frac{\partial u}{\partial x}\right)_{i,m} + \left(\frac{\partial u}{\partial t}\right)_{i,m} + \mathcal{O}(\Delta x^2)\end{aligned} \tag{4-23}$$

Now:
$$u_{i,n+1} = u_{i,m} + (1-\theta)\Delta t \left(\frac{\partial u}{\partial t}\right)_{i,m} + \tfrac{1}{2}(1-\theta)^2 \Delta t^2 \left(\frac{\partial^2 u}{\partial t^2}\right)_{i,m} + \mathcal{O}(\Delta t^3)$$

$$u_{i,n} = u_{i,m} - \theta \Delta t \left(\frac{\partial u}{\partial t}\right)_{i,m} + \tfrac{1}{2}\theta^2 \Delta t^2 \left(\frac{\partial^2 u}{\partial t^2}\right)_{i,m} + \mathcal{O}(\Delta t^3)$$

so that:

$$\frac{u_{i,n+1} - u_{i,n}}{\Delta t} = \left(\frac{\partial u}{\partial t}\right)_{i,m} + (\tfrac{1}{2} - \theta)\Delta t \left(\frac{\partial^2 u}{\partial t^2}\right)_{i,m} + \mathcal{O}(\Delta t^2) \qquad (4\text{-}24)$$

In addition:

$$\left(\frac{\partial u}{\partial x}\right)_{i,n+1} = \left(\frac{\partial u}{\partial x}\right)_{i,m} + (1-\theta)\Delta t \left(\frac{\partial^2 u}{\partial t \partial x}\right)_{i,m} + \mathcal{O}(\Delta t^2)$$

$$\left(\frac{\partial u}{\partial x}\right)_{i,n} = \left(\frac{\partial u}{\partial x}\right)_{i,m} - \theta \Delta t \left(\frac{\partial^2 u}{\partial t \partial x}\right)_{i,m} + \mathcal{O}(\Delta t^2)$$

so that:

$$\theta \left(\frac{\partial u}{\partial x}\right)_{i,n+1} + (1-\theta)\left(\frac{\partial u}{\partial x}\right)_{i,n} = \left(\frac{\partial u}{\partial x}\right)_{i,m} + \mathcal{O}(\Delta t^2) \qquad (4\text{-}25)$$

Similarly:

$$\theta \left(\frac{\partial^2 u}{\partial x^2}\right)_{i,n+1} + (1-\theta)\left(\frac{\partial^2 u}{\partial x^2}\right)_{i,n} = \left(\frac{\partial^2 u}{\partial x^2}\right)_{i,m} + \mathcal{O}(\Delta t^2) \qquad (4\text{-}26)$$

Substitution of eqs. (4-24), (4-25), and (4-26) into eq. (4-23) and cancellation of common terms yields:

$$\epsilon_L = -vf'\Delta x(\tfrac{1}{2} - W)\left(\frac{\partial^2 u}{\partial x^2}\right)_{i,m} + (\theta - \tfrac{1}{2})\Delta t \left(\frac{\partial^2 u}{\partial t^2}\right)_{i,m} + \mathcal{O}(\Delta x^2) + \mathcal{O}(\Delta t^2) \qquad (4\text{-}27)$$

Now, differentiation of eq. (4-21) with respect to t gives:

$$\frac{\partial^2 u}{\partial t^2} = -vf' \frac{\partial^2 u}{\partial t \partial x} = -vf' \frac{\partial^2 u}{\partial x \partial t}$$

while differentiation with respect to x gives:

$$\frac{\partial^2 u}{\partial x \partial t} = -vf' \frac{\partial^2 u}{\partial x^2}$$

so that:

$$\frac{\partial^2 u}{\partial t^2} = (vf')^2 \frac{\partial^2 u}{\partial x^2}$$

Substitution into eq. (4-27) gives the final form for the local truncation error:

$$\epsilon_L = vf'\Delta x[(W-\tfrac{1}{2}) + \lambda(\theta-\tfrac{1}{2})]\left(\frac{\partial^2 u}{\partial x^2}\right)_{i,m} + \mathcal{O}(\Delta x^2) + \mathcal{O}(\Delta t^2) \qquad (4\text{-}28)$$

Numerical dispersion

Equation (2-23) can be written in the form:

$$L_D u_{i,m} = L u_{i,m} + \epsilon_L = -vf'\frac{\partial u}{\partial x} - \frac{\partial u}{\partial t} + \epsilon_L$$

Upon substituting eq. (4-28), we have:

$$L_D u_{i,m} = \mathcal{D}_{num}\frac{\partial^2 u}{\partial x^2} - vf'\frac{\partial u}{\partial x} - \frac{\partial u}{\partial t} + \mathcal{O}(\Delta x^2) + \mathcal{O}(\Delta t^2)$$

where:

$$\mathcal{D}_{num} = vf'\Delta x[(W-\tfrac{1}{2}) + \lambda(\theta-\tfrac{1}{2})] \qquad (4\text{-}29)$$

Thus, by solving the difference equation (4-11), we are, in effect, solving the diffusion-convection problem:

$$\mathcal{D}_{num}\frac{\partial^2 u}{\partial x^2} - vf'\frac{\partial u}{\partial x} = \frac{\partial u}{\partial t} \qquad (4\text{-}30)$$

(a one-dimensional equivalent to eq. (1-64)), rather than the pure convection problem expressed by eq. (4-21). The truncation error associated with the use of the various difference equations discussed above has therefore been described by the term *numerical dispersion*.

In Table IV, $\mathcal{D}_{num}$ is listed for each of the nine special difference equations. It is interesting to note that in every case, positive $\mathcal{D}_{num}$ is associated with stability, while negative $\mathcal{D}_{num}$ is associated with instability. In only two cases is $\mathcal{D}_{num} = 0$, i.e.:

(a) Forward-in-time, backward-in-distance ($W = 1$, $\theta = 0$), with λ *exactly* equal to one.

(b) Centered-in-time, centered-in-distance ($W = \theta = \tfrac{1}{2}$).

In both of these cases, neutral stability obtains, wherein $|\gamma| = 1$.

Thus, it is clear that there is a close connection between stability and numerical dispersion and that, in effect, the numerical dispersion acts to stabilize the difference equation.

Superposition of numerical and physical dispersion

If we consider solving the diffusion-convection equation with a "physical" diffusivity, $\mathcal{D}$:

$$\mathcal{D}\frac{\partial^2 u}{\partial x^2} - vf'\frac{\partial u}{\partial x} = \frac{\partial u}{\partial t} \tag{4-31}$$

and derive a difference analog in the same manner as discussed above for the convection equation, then truncation error analysis will show that the solution of the difference equation corresponds to the solution of the modified problem:

$$(\mathcal{D} + \mathcal{D}_{num})\frac{\partial^2 u}{\partial x^2} - vf'\frac{\partial u}{\partial x} = \frac{\partial u}{\partial t} \tag{4-32}$$

A serious problem now becomes evident. When the physical diffusivity, $\mathcal{D}$, is small, the true solution should be a relatively sharp front, and it may be important to obtain from the calculation the "width" of the front. When numerical dispersion is severe, as it so frequently is (unfortunately), $\mathcal{D}_{num}$ can be much larger than $\mathcal{D}$. As a result, numerical dispersion will swamp physical dispersion, leading to a front apparently much more smeared than it should be.

Equation (4-29) suggests that numerical dispersion can be reduced by reducing Δx, but reduction to a tolerable level may require an excessive number of grid points. It can be controlled to some extent by adjustment of the weighting factor, W, but care must be taken not to let the difference equation go unstable, which can lead to an oscillatory or otherwise unrealistic solution.

EXAMPLE CALCULATIONS

Purpose and details of calculations

To illustrate some of the ideas developed above, we consider some examples of the numerical solution of the diffusion-convection equation (4-31). These solutions are all obtained by use of the difference equation (4-11), modified by adding to the left-hand side the following centered-in-distance, centered-in-time analog of the "physical" diffusion term:

$$(\mathcal{D}\Delta t/2\Delta x^2)(U_{i+1,n+1} - 2U_{i,n+1} + U_{i-1,n+1} + U_{i+1,n} - 2U_{i,n} + U_{i-1,n})$$

Note that the weighting factors, W and θ, are not applied to the diffusion term, but only to the convection term.

For the initial condition:

$$u(x, 0) = 0, \quad x > 0 \tag{4-33}$$

and boundary conditions:

$$u(0, t) = 1, \quad u(\infty, t) = 0, \quad t > 0 \tag{4-34}$$

eq. (4-31) has the exact solution:

$$u = \tfrac{1}{2} \operatorname{erfc}\left[\frac{x - vf't}{2(\mathcal{D}t)^{\frac{1}{2}}}\right] + \tfrac{1}{2} \exp(vf'x/\mathcal{D}) \operatorname{erfc}\left[\frac{x + vf't}{2(\mathcal{D}t)^{\frac{1}{2}}}\right] \tag{4-35}$$

where erfc is the complementary error function, defined by:

$$\operatorname{erfc}(x) = 1 - (2/\pi^{\frac{1}{2}}) \int_0^x \exp(-u^2)\,du$$

The corresponding initial and boundary conditions for the numerical examples are:

$$U_{i,0} = 0, \quad i \geq 1 \tag{4-36}$$

$$U_{0,n} = 1, \quad U_{I,n} = 0, \quad n \geq 1 \tag{4-37}$$

The last equation is an adequate representation of the boundary condition at $x = \infty$ if I is chosen sufficiently large that the solution does not change appreciably at the point $i = I - 1$. In the example calculations, $I = 5/\Delta x$ (corresponding to $x = 5$) satisfies this condition.

The exact solution (4-35) is indeterminate at the point $x = 0$, $t = 0$. However, the numerical procedure (in most cases) requires a value for $U_{0,0}$. For the example calculations, the arbitrary choice of $U_{0,0} = 0.5$ is made.

Further specifics of the calculation are:

$\Delta x = 0.1$

$vf' = 1.0$

$\mathcal{D} = 0.01$

$\left.\begin{array}{l}\Delta t = 0.05 \\ \lambda = 0.5\end{array}\right\}$ or $\left\{\begin{array}{l}\Delta t = 0.1 \\ \lambda = 1.0\end{array}\right.$

At each time step, solution is accomplished by use of the tridiagonal algorithm described in Chapter 3.

Solutions showing numerical dispersion

Figures 10 and 11 show six examples of solutions that exhibit significant numerical dispersion. In each example, the exact solution to eq. (4-31), namely eq. (4-35) for $\mathcal{D} = 0.01$ and $t = 1.0$, is plotted as the solid curve. Corresponding numerical values for $\mathcal{D}_{num}$, obtained from eq. (4-29), are noted in the legends. The numerical solutions exhibit significantly more smearing than the exact solutions, and the amount of smearing clearly increases as $\mathcal{D}_{num}$ increases. For comparison, eq. (4-35) has also been evaluated using a "total" diffusion coefficient:

$$\mathcal{D}_t = \mathcal{D} + \mathcal{D}_{num}$$

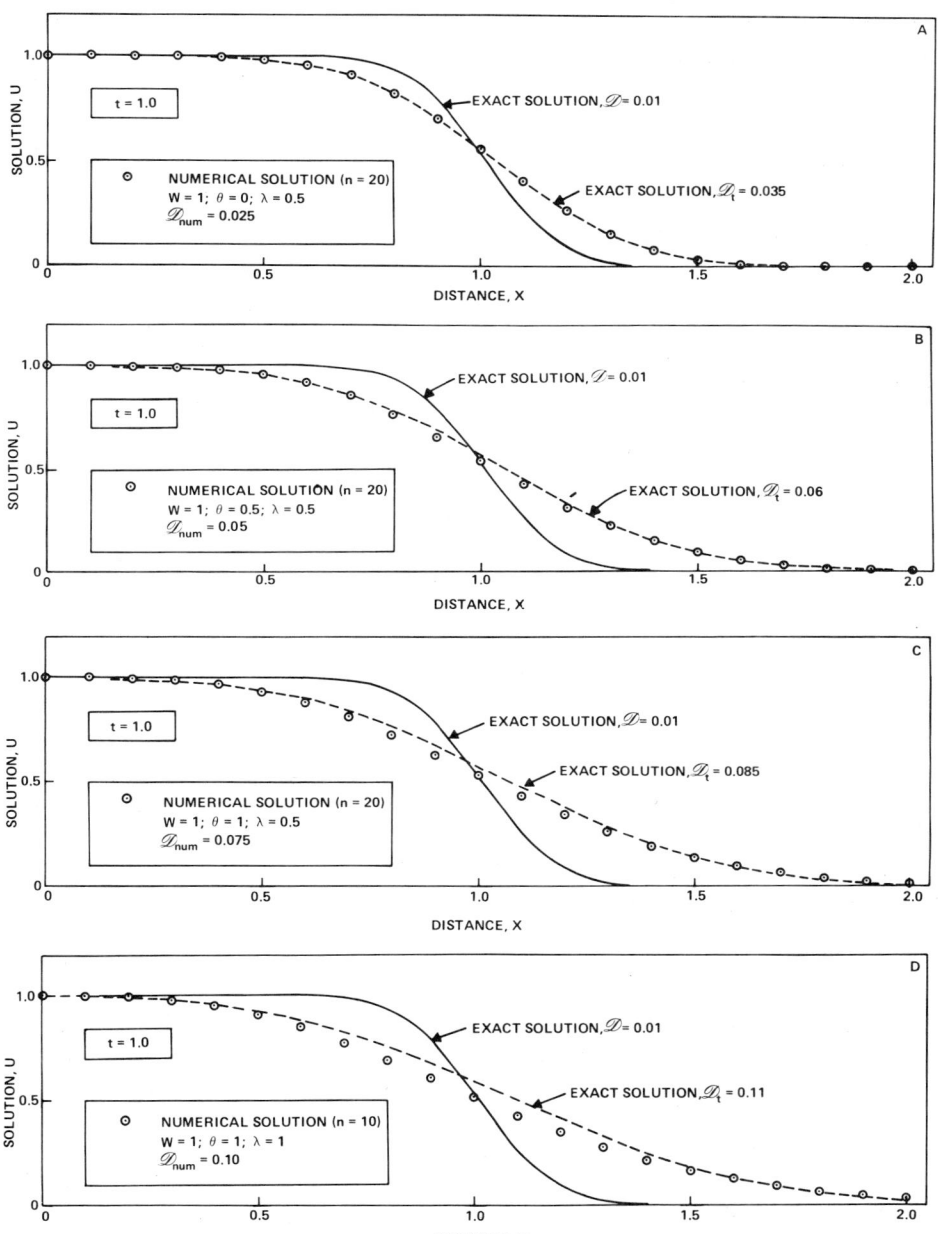

Fig. 10. Numerical solutions (of diffusion-convection equation) exhibiting numerical dispersion. Difference equations are backward-in-distance and: A. Forward-in-time, with $\lambda = \frac{1}{2}$. B. Centered-in-time, with $\lambda = \frac{1}{2}$. C. Backward-in-time, with $\lambda = \frac{1}{2}$. D. Backward-in-time, with $\lambda = 1$.

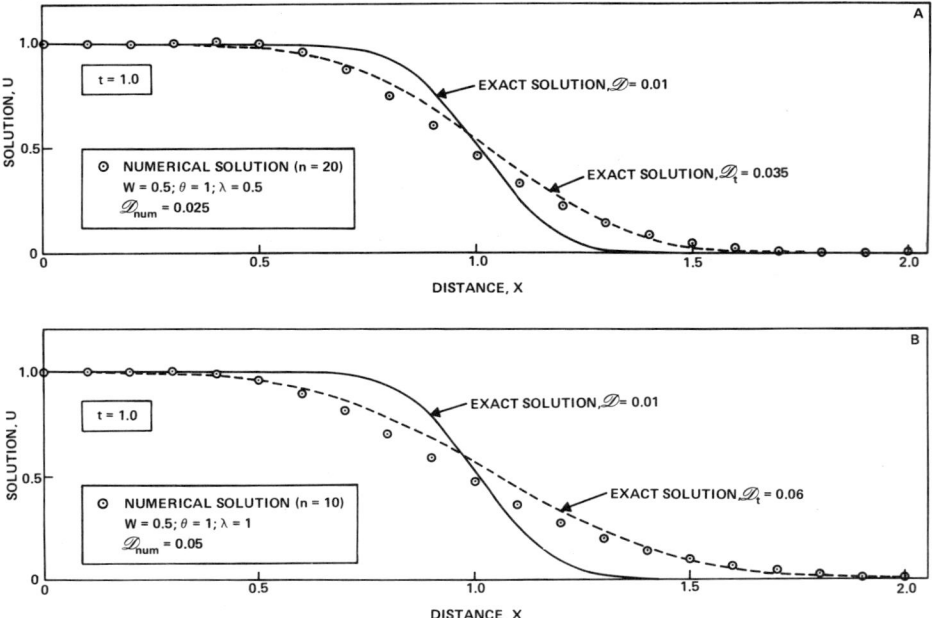

Fig. 11. Numerical solutions (of diffusion-convection equation) exhibiting numerical dispersion. Difference equations are centered-in-distance and backward-in-time. A. $\lambda = \tfrac{1}{2}$. B. $\lambda = 1$.

and this is shown as the dashed curve in each figure. Good-to-excellent agreement exists between the numerical solutions and the "exact" solutions of eq. (4-35) that use the "total" diffusion coefficient.

Solutions without numerical dispersion

Figure 12 shows solutions for three cases where $\mathcal{D}_{num} = 0$. Case A involves the backward-in-distance, forward-in-time difference equation, for which zero numerical dispersion can be obtained only by choosing the time step, Δt, so that $\lambda = 1$. This choice of λ yields excellent agreement between the numerical and exact solutions. Indeed, if vf' is constant and the size of the time step is freely at one's disposal, such an approach to avoiding numerical dispersion is highly recommended. Unfortunately, this approach cannot be carried over to more complex situations of interest, such as multiphase problems where f' is not constant, or to multidimensional problems where v_x and v_y are not constant.

Cases B and C in Fig. 12 involve another approach to reducing $\mathcal{D}_{num}$ to zero, namely the use of a centered-in-distance, centered-in-time difference equation. Here, the numerical solutions oscillate somewhat, with overshoot occurring behind the moving front. This overshoot is typical of many higher-

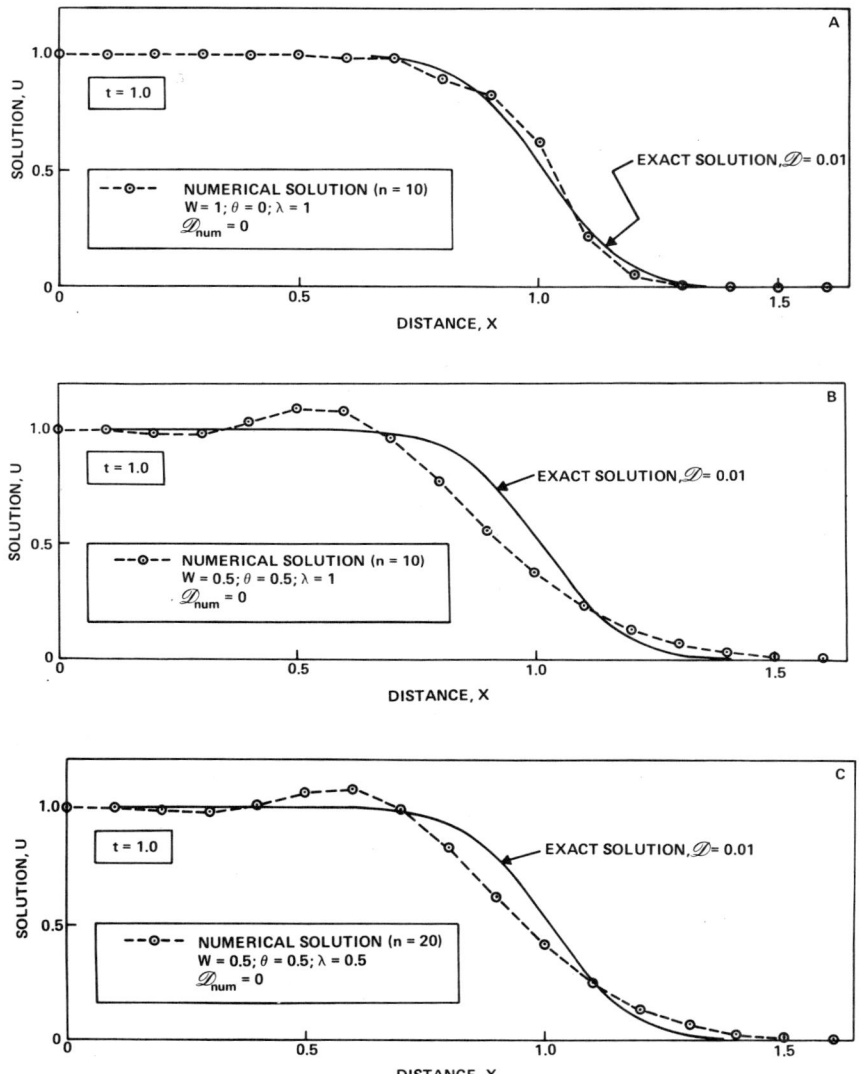

Fig. 12. Numerical solutions (of diffusion-convection equation) with zero numerical dispersion. Difference equations are: A. Backward-in-distance, forward-in-time, with $\lambda = 1$. B. Centered-in-distance, centered-in-time, with $\lambda = 1$. C. Centered-in-distance, centered-in-time, with $\lambda = \frac{1}{2}$.

order difference equations designed to eliminate numerical dispersion, and the overshoot tends to become more severe as the front becomes sharper (i.e., as the physical diffusion coefficient, $\mathscr{D}$, becomes smaller).

Unstable solutions

Finally, Fig. 13 shows two examples of unstable numerical solutions obtained with a centered-in-distance, forward-in-time difference equation. In both examples, the numerical dispersion coefficient (as well as the "total" diffusion coefficient, $\mathscr{D}_t = \mathscr{D} + \mathscr{D}_{num}$) is negative, and the degree of insta-

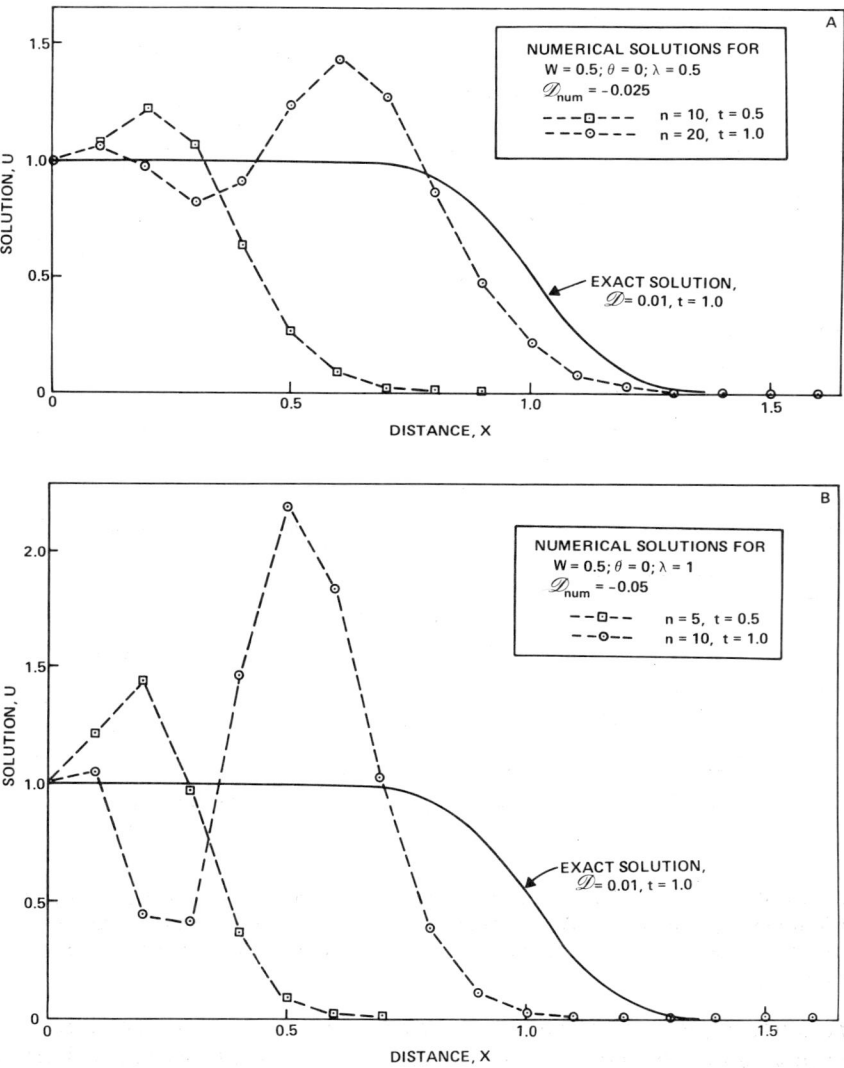

Fig. 13. Numerical solutions (of diffusion-convection equation) exhibiting instability. Difference equations are centered-in-distance, forward-in-time. A. $\lambda = \frac{1}{2}$. B. $\lambda = 1$.

bility (i.e., growth of the oscillation with each time step) increases as λ and $-\mathcal{D}_{num}$ increase.

One can also carry out calculations for the forward-in-distance difference equations (i.e., $W = 0$), but the instability is so great as to preclude plotting the solutions in any meaningful way.

SUMMARY

Example calculations clearly indicate that a numerical solution of the diffusion-convection equation can be expected to exhibit either numerical dispersion or some oscillatory behavior. (The one exception, illustrated by Fig. 12A, must be regarded as a special case, since it cannot be generalized to more complex situations.) While the above examples and truncation analyses apply only to the linearized version of the diffusion-convection equation, the same phenomena can be observed for nonlinear versions associated with multiphase flow. The influence of the nonlinearity is not completely understood, so quantitative interpretation of numerical dispersion or oscillatory behavior is, at this time, somewhat elusive. Lantz (1971) does provide some guidelines, however.

In any event, the user of a reservoir simulator needs to be aware that either numerical dispersion or oscillatory behavior (usually in the form of overshoot) can be expected in convection-dominated problems and that he can, to some extent, trade off between one or the other by adjusting the weighting factors. But he cannot make them both disappear. His primary recourse is to refine the grid, but this, of course, is limited by practical considerations involving computational costs as well as available computing time and storage.

The subject of this chapter is an active area of research, and several higher-order difference equations for solving the diffusion-convection equation have been proposed. They include the two-point upstream weighting method of Todd et al. (1972), the variational approach of Price et al. (1968), higher-order difference equations of Stone and Brian (1963) and of Chaudhari (1971), and, most recently, a new "truncation-cancellation" procedure of Laumbach (1975). A detailed discussion of these equations is beyond the scope of this book. While some look very promising, at least for the linearized convection-diffusion problem, only a few have been implemented in existing reservoir simulators, and the successful application of the others to multiphase flow problems has yet to be demonstrated.

NOMENCLATURE

Symbols

$\mathcal{D}$	=	physical diffusivity [m²/s]
$\mathcal{D}_{num}$	=	numerical diffusivity [m²/s]
$\mathcal{D}_t$	=	$\mathcal{D} + \mathcal{D}_{num}$ [m²/s]
f	=	function of dependent variable u
f'	=	df/du
F	=	difference solution for function f
i	=	$(-1)^{\frac{1}{2}}$
I	=	number of grid points in x-direction
t	=	time [s]
u	=	general function
U	=	difference solution for function u
v	=	velocity [m/s]
W	=	distance-weighting parameter
x	=	distance [m]
Δt	=	increment of t [s]
Δx	=	increment of x [m]
θ	=	time-weighting parameter
λ	=	$vf'\Delta t/\Delta x$

Subscripts

i	index in x-direction
m	fractional index in t-direction $= (1 - \theta)n + \theta(n + 1)$
n	index in t-direction (i.e., time)

CHAPTER 5

NUMERICAL SOLUTION OF ELLIPTIC PROBLEMS IN ONE DEPENDENT VARIABLE

ELLIPTIC DIFFERENCE EQUATIONS

Formulation

Consider the two-dimensional elliptic differential equation:

$$\frac{\partial}{\partial x}\left(KX\frac{\partial p}{\partial x}\right) + \frac{\partial}{\partial y}\left(KY\frac{\partial p}{\partial y}\right) + q(x, y, p) = 0,$$

$$0 < x < L, \quad 0 < y < W \tag{5-1}$$

with "Neumann" boundary conditions:

$$\frac{\partial p}{\partial x} = 0 \text{ at } x = 0 \text{ and } x = L$$

$$\frac{\partial p}{\partial y} = 0 \text{ at } y = 0 \text{ and } y = W \tag{5-2}$$

Equation (5-1) is sufficiently general to represent two-dimensional incompressible flow of a single fluid in a nonhomogeneous anisotropic medium, where $q(x, y, p)$ represents distributed injection (if positive) or production (if negative).

For further generality, we will set up a grid with variable spacing; i.e., Δx will vary with x and Δy will vary with y. As usual, we will let i be the index in the x-direction and j be the index in the y-direction.

It was pointed out in Chapter 2 that two types of grids are in common use in reservoir work, but that the difference equation is written the same way for either grid. The derivatives are approximated by:

$$\frac{\partial}{\partial x}\left(KX\frac{\partial p}{\partial x}\right) \approx \frac{KX_{i+\frac{1}{2},j}\dfrac{P_{i+1,j} - P_{ij}}{x_{i+1} - x_i} - KX_{i-\frac{1}{2},j}\dfrac{P_{ij} - P_{i-1,j}}{x_i - x_{i-1}}}{x_{i+\frac{1}{2}} - x_{i-\frac{1}{2}}} \tag{5-3}$$

$$\frac{\partial}{\partial y}\left(KY\frac{\partial p}{\partial y}\right) \approx \frac{KY_{i,j+\frac{1}{2}}\dfrac{P_{i,j+1} - P_{ij}}{y_{j+1} - y_j} - KY_{i,j-\frac{1}{2}}\dfrac{P_{ij} - P_{i,j-1}}{y_j - y_{j-1}}}{y_{j+\frac{1}{2}} - y_{j-\frac{1}{2}}} \tag{5-4}$$

Let us make the following definitions:

$$\Delta x_i = x_{i+\frac{1}{2}} - x_{i-\frac{1}{2}}$$

$$\Delta y_j = y_{j+\frac{1}{2}} - y_{j-\frac{1}{2}}$$

$$AX_{i+\frac{1}{2},j} = \frac{KX_{i+\frac{1}{2},j}\Delta y_j}{x_{i+1} - x_i} \equiv \text{transmissibility in } \hat{\imath}$$

$$AY_{i,j+\frac{1}{2}} = \frac{KY_{i,j+\frac{1}{2}}\Delta x_i}{y_{j+1} - y_j}$$

$$Q_{ij} = \Delta x_i \Delta y_j q(x_i, y_j, P_{ij})$$

Then, if we substitute eqs. (5-3) and (5-4) into eq. (5-1), multiply through by $\Delta x_i \, \Delta y_j$, and change signs, we obtain the following:

$$-AX_{i+\frac{1}{2},j}(P_{i+1,j} - P_{ij}) + AX_{i-\frac{1}{2},j}(P_{ij} - P_{i-1,j}) - AY_{i,j+\frac{1}{2}}(P_{i,j+1} - P_{ij})$$
$$+ AY_{i,j-\frac{1}{2}}(P_{ij} - P_{i,j-1}) = Q_{ij} \tag{5-5}$$

Equation (5-5) is applied at every grid point in the solution rectangle. For some points (either near or on the boundary), it will involve one or two fictitious points outside the rectangle; values of P at these points are eliminated by use of eq. (2-17) or (2-18), depending on which type of grid is being used. We thus have a system of $I \cdot J$ equations with $I \cdot J$ unknowns, P_{ij}.

Q_{ij} may be interpreted as the integral of $q(x, y, p)$ over the area $\Delta x_i \, \Delta y_j$; normally, however, it represents an injection rate (production rate, if negative) of a well located at the point (x_i, y_j).

Equations of the form (5-5) also arise at each time step in the solution of two-dimensional parabolic problems by an implicit difference equation, such as (3-40) or (3-42). In this case, Q_{ij} will involve values of the variable at the known time level, t_n.

Matrix notation; the structure of the coefficient matrix

Many practical problems in a wide variety of fields give rise to linear systems of equations of the form:

$$a_{1,1}y_1 + a_{1,2}y_2 + \ldots + a_{1,N}y_N = k_1$$
$$a_{2,1}y_1 + a_{2,2}y_2 + \ldots + a_{2,N}y_N = k_2$$
$$\vdots \qquad \vdots$$
$$a_{N,1}y_1 + a_{N,2}y_2 + \ldots + a_{N,N}y_N = k_N \tag{5-6}$$

where N is the number of equations and the number of unknowns. The system can be concisely represented in matrix form as:

$$\mathbf{Ay} = \mathbf{k} \tag{5-7}$$

$$\begin{bmatrix} a_{1,1} & a_{1,2} & \cdots & a_{1,N} \\ a_{2,1} & a_{2,2} & \cdots & a_{2,N} \\ \vdots & & & \vdots \\ a_{N,1} & a_{N,2} & \cdots & a_{N,N} \end{bmatrix}$$

Fig. 14. General matrix, A.

where **y** is the unknown column vector $(y_1, y_2, \ldots, y_N)$, **k** is the known column vector $(k_1, k_2, \ldots, k_N)$, and **A** is the matrix shown in Fig. 14. Standard matrix rules for multiplying a matrix by a column vector give the equivalence of eqs. (5-7) and (5-6).

Identify the elements of **A** by a_{rs}, where r is the row index and s is the column index. The main diagonal of **A** consists of those elements for which $r = s$. The matrix **A** is symmetric if $a_{rs} = a_{sr}$; i.e., if elements below the main diagonal are equal to corresponding (or reflected) elements above the main diagonal.

A sparse matrix is one in which most of the elements are zero. Finite-difference methods generally yield sparse matrices, which allow special techniques to be used for their solution.

An upper triangular matrix is one in which all the elements below and to the left of the main diagonal are zero; a lower triangular matrix is one in which all the elements above and to the right of the main diagonal are zero. A diagonal matrix has zero elements on all but the main diagonal.

Consider the difference equation (5-5) with boundary conditions (2-17). Let us choose $I = 4$, $J = 3$. In order to apply matrix methods, we must choose an ordering of the points (i, j). Let us pick the ordering shown in Fig. 15. The complete set of equations is shown in Fig. 16.

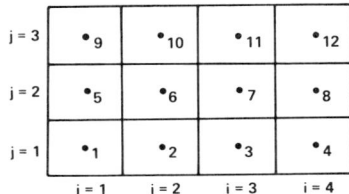

Fig. 15. Computing rectangle divided into numbered blocks.

In Fig. 16, the structure is not apparent. Let us examine the equations in matrix form, shown in Fig. 17. The diagonal elements are:

$$d_{1,1} = AX_{1\frac{1}{2},1} + AY_{1,1\frac{1}{2}}$$
$$d_{2,1} = AX_{1\frac{1}{2},1} + AX_{2\frac{1}{2},1} + AY_{2,1\frac{1}{2}}$$
$$d_{3,1} = AX_{2\frac{1}{2},1} + AX_{3\frac{1}{2},1} + AY_{3,1\frac{1}{2}}$$

$$-AX_{1\frac{1}{2},1}(P_2 - P_1) \qquad -AY_{1,1\frac{1}{2}}(P_5 - P_1) \qquad = Q_{1,1}$$

$$-AX_{2\frac{1}{2},1}(P_3 - P_2) + AX_{1\frac{1}{2},1}(P_2 - P_1) - AY_{2,1\frac{1}{2}}(P_6 - P_2) \qquad = Q_{2,1}$$

$$-AX_{3\frac{1}{2},1}(P_4 - P_3) + AX_{2\frac{1}{2},1}(P_3 - P_2) - AY_{3,1\frac{1}{2}}(P_7 - P_3) \qquad = Q_{3,1}$$

$$AX_{3\frac{1}{2},1}(P_4 - P_3) - AY_{4,1\frac{1}{2}}(P_8 - P_4) \qquad = Q_{4,1}$$

$$-AX_{1\frac{1}{2},2}(P_6 - P_5) \qquad -AY_{1,2\frac{1}{2}}(P_9 - P_5) + AY_{1,1\frac{1}{2}}(P_5 - P_1) = Q_{1,2}$$

$$-AX_{2\frac{1}{2},2}(P_7 - P_6) + AX_{1\frac{1}{2},2}(P_6 - P_5) - AY_{2,2\frac{1}{2}}(P_{10} - P_6) + AY_{2,1\frac{1}{2}}(P_6 - P_2) = Q_{2,2}$$

$$-AX_{3\frac{1}{2},2}(P_8 - P_7) + AX_{2\frac{1}{2},2}(P_7 - P_6) - AY_{3,2\frac{1}{2}}(P_{11} - P_7) + AY_{3,1\frac{1}{2}}(P_7 - P_3) = Q_{3,2}$$

$$AX_{3\frac{1}{2},2}(P_8 - P_7) - AY_{4,2\frac{1}{2}}(P_{12} - P_8) + AY_{4,1\frac{1}{2}}(P_8 - P_4) = Q_{4,2}$$

$$-AX_{1\frac{1}{2},3}(P_{10} - P_9) \qquad + AY_{1,2\frac{1}{2}}(P_9 - P_5) = Q_{1,3}$$

$$-AX_{2\frac{1}{2},3}(P_{11} - P_{10}) + AX_{1\frac{1}{2},3}(P_{10} - P_9) \qquad + AY_{2,2\frac{1}{2}}(P_{10} - P_6) = Q_{2,3}$$

$$-AX_{3\frac{1}{2},3}(P_{12} - P_{11}) + AX_{2\frac{1}{2},3}(P_{11} - P_{10}) \qquad + AY_{3,2\frac{1}{2}}(P_{11} - P_7) = Q_{3,3}$$

$$AX_{3\frac{1}{2},3}(P_{12} - P_{11}) \qquad + AY_{4,2\frac{1}{2}}(P_{12} - P_8) = Q_{4,3}$$

Fig. 16. Equations for $I = 4$, $J = 3$.

$$d_{4,1} = AX_{3\frac{1}{2},1} + AY_{4,1\frac{1}{2}}$$
$$d_{1,2} = AX_{1\frac{1}{2},2} + AY_{1,1\frac{1}{2}} + AY_{1,2\frac{1}{2}}$$
$$d_{2,2} = AX_{1\frac{1}{2},2} + AX_{2\frac{1}{2},2} + AY_{2,1\frac{1}{2}} + AY_{2,2\frac{1}{2}}$$
$$d_{3,2} = AX_{2\frac{1}{2},2} + AX_{3\frac{1}{2},2} + AY_{3,1\frac{1}{2}} + AY_{3,2\frac{1}{2}}$$
$$d_{4,2} = AX_{3\frac{1}{2},2} + AY_{4,1\frac{1}{2}} + AY_{4,2\frac{1}{2}}$$
$$d_{1,3} = AX_{1\frac{1}{2},3} + AY_{1,2\frac{1}{2}}$$
$$d_{2,3} = AX_{1\frac{1}{2},3} + AX_{2\frac{1}{2},3} + AY_{2,2\frac{1}{2}}$$
$$d_{3,3} = AX_{2\frac{1}{2},3} + AX_{3\frac{1}{2},3} + AY_{3,2\frac{1}{2}}$$
$$d_{4,3} = AX_{3\frac{1}{2},3} + AY_{4,2\frac{1}{2}}$$

Now the structure becomes apparent. The coefficient matrix, of order $N = I \cdot J$, can be partitioned into $J \times J$ submatrices, where each submatrix is of order $I \times I$. Thus, A has the "block" triangular form shown in Fig. 18, where $D_{j+\frac{1}{2}}$ is a diagonal matrix of order $I \times I$, and whose diagonal elements are $-AY_{i,j+\frac{1}{2}}$; and T_j is a tridiagonal matrix of order $I \times I$ of the form shown in Fig. 19.

We note that this coefficient matrix is symmetric. In addition, the diagonal elements d_{ij} are found to be equal to the negative sum of the off-diagonal elements on each row. Finally, we note that the matrix is sparse, particularly for large I and J. In particular, this matrix is a special case of a "band matrix". A band matrix has all elements that are outside of a band of width W centered on the main diagonal equal to zero. Thus:

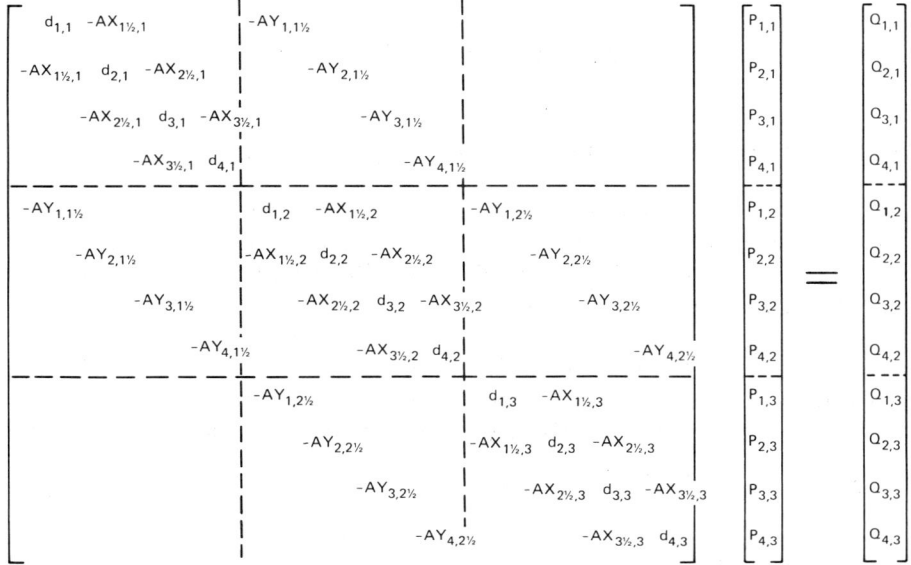

Fig. 17. Equations in matrix form for $I = 4$, $J = 3$.

$$A = \begin{bmatrix} T_1 & D_{1\frac{1}{2}} & & & & \\ D_{1\frac{1}{2}} & T_2 & D_{2\frac{1}{2}} & & & \\ & D_{2\frac{1}{2}} & T_3 & D_{3\frac{1}{2}} & & \\ & & & \ddots & & \\ & & & & D_{J-\frac{1}{2}} & T_J \end{bmatrix}$$

Fig. 18. Coefficient matrix in block-triangular form, where each element is a submatrix.

$$T_j = \begin{bmatrix} d_{1,j} & -AX_{1\frac{1}{2},j} & & & & \\ -AX_{1\frac{1}{2},j} & d_{2,j} & -AX_{2\frac{1}{2},j} & & & \\ & -AX_{2\frac{1}{2},j} & d_{3,j} & -AX_{3\frac{1}{2},j} & & \\ & & & \ddots & & \\ & & & & -AX_{I-\frac{1}{2},j} & d_{I,j} \end{bmatrix}$$

Fig. 19. Tridiagonal matrix, T_j.

$$a_{r,s} = 0 \begin{cases} \text{for } r < s - \tfrac{1}{2}(W-1) \\ \text{or for } r > s + \tfrac{1}{2}(W-1) \end{cases}$$

DIRECT SOLUTION OF BAND MATRIX EQUATIONS BY FACTORIZATION

In Chapter 3, we encountered tridiagonal equations, which are band equations of width $W = 3$. Band equations of larger width arise, as we have just seen, in the solution of two-dimensional elliptic and parabolic problems. We shall see later that they also arise in the solution of certain types of two-phase flow problems.

We shall present here a general algorithm for the solution of band equations by the method of factorization. The band matrix **B** of order $N \times N$ and width W will be factored into lower and upper triangular band matrices:

$$\mathbf{B} = \mathbf{LU} \tag{5-8}$$

There are N degrees of freedom in the factorization; we shall choose **U** in such a way that all the elements on its main diagonal are *ones*. Once the factorization is performed, we can solve:

$$\mathbf{Bp} = \mathbf{q} \tag{5-9}$$

as follows. Write:

$$\mathbf{LUp} = \mathbf{q} \tag{5-10}$$

Let $\mathbf{g} = \mathbf{Up}$. Then:

$$\mathbf{Lg} = \mathbf{q} \tag{5-11}$$

$$\mathbf{Up} = \mathbf{g} \tag{5-12}$$

Since **L** and **U** are triangular, their solution is quite straightforward. Equation (5-11) is referred to as the forward solution; (5-12) is the back solution.

First let us derive an algorithm for factoring any *full* matrix. Let b_{rs} be the elements of **B**, l_{rs} the elements of **L**, and u_{rs} the elements of **U**. Then the matrix-multiplication algorithm corresponding to eq. (5-8) is:

$$b_{rs} = \sum_{k=1}^{N} l_{rk} u_{ks} \tag{5-13}$$

We work on the rows in the sequence $r = 1, 2, \ldots, N$. For each row r, we solve for the elements l_{rs}, $s = 1, 2, \ldots, r$; we then solve for the elements u_{rs}, $s = r+1, \ldots, N$. First break up the sum in eq. (5-13) as follows:

$$\sum_{k=1}^{s-1} l_{rk} u_{ks} + l_{rs} u_{ss} + \sum_{k=s+1}^{N} l_{rk} u_{ks} = b_{rs}$$

In the last sum $u_{ks} = 0$ since $k > s$ and u_{ks} is an element of an upper triangular matrix. Further, in the second term, $u_{ss} = 1$. Thus we have:

$$l_{rs} = b_{rs} - \sum_{k=1}^{s-1} l_{rk} u_{ks} \quad \begin{cases} r = 1, 2, \ldots, N \\ s = 1, 2, \ldots, r \end{cases} \tag{5-14}$$

Now let us break up the sum in eq. (5-13) as follows:

$$\sum_{k=1}^{r-1} l_{rk} u_{ks} + l_{rr} u_{rs} + \sum_{k=r+1}^{N} l_{rk} u_{ks} = b_{rs}$$

In the last sum, $l_{rk} = 0$ since $k > r$ and l_{rk} is an element of a lower triangular matrix. Thus:

$$u_{rs} = \frac{b_{rs} - \sum_{k=1}^{r-1} l_{rk} u_{ks}}{l_{rr}} \quad \begin{cases} r = 1, 2, \ldots, N \\ s = r+1, \ldots, N \end{cases} \tag{5-15}$$

The matrix-by-vector-multiplication algorithm corresponding to the forward solution (5-11) is:

$$\sum_{k=1}^{N} l_{rk} g_k = q_r$$

The sum may be broken up as follows:

$$\sum_{k=1}^{r-1} l_{rk} g_k + l_{rr} g_r + \sum_{k=r+1}^{N} l_{rk} g_k = q_r$$

But in the last sum, $l_{rk} = 0$. Hence:

$$g_r = \frac{q_r - \sum_{k=1}^{r-1} l_{rk} g_k}{l_{rr}}, \quad r = 1, 2, \ldots, N \tag{5-16}$$

The matrix-by-vector-multiplication algorithm corresponding to the back solution (5-12) is:

$$\sum_{k=1}^{N} u_{rk} p_k = g_r$$

This sum may be broken up as:

$$\sum_{k=1}^{r-1} u_{rk} p_k + u_{rr} p_r + \sum_{k=r+1}^{N} u_{rk} p_k = g_r$$

In the first sum, $u_{rk} = 0$, since $k < r$. Also $u_{rr} = 1$. Thus:

$$p_r = g_r - \sum_{k=r+1}^{N} u_{rk} p_k, \quad r = N, N-1, \ldots, 1 \tag{5-17}$$

Note that p_r *must* be evaluated in reverse sequence.

We now specialize these algorithms for the case of a band matrix. Let $W' = (W-1)/2$, and:

$L_1(r) = \max(1, r - W')$

$L_2(r) = \min(N, r + W')$

$L_3(s) = \max(1, s - W')$

Then the algorithms (5-14) to (5-17) become:
Factorization: $r = 1, 2, \ldots, N.$

$$l_{rs} = b_{rs} - \sum_{k=L_1(r)}^{s-1} l_{rk} u_{ks} \qquad s = L_1, L_1+1, \ldots, r \tag{5-18}$$

$$u_{rs} = \left(b_{rs} - \sum_{k=L_3(s)}^{r-1} l_{rk} u_{ks}\right) \Big/ l_{rr} \qquad s = r+1, \ldots, L_2 \tag{5-19}$$

Forward solution: $r = 1, 2, \ldots, N.$

$$g_r = \left(q_r - \sum_{k=L_1(r)}^{r-1} l_{rk} g_k\right) \Big/ l_{rr} \tag{5-20}$$

Back solution: $r = N, N-1, \ldots, 1.$

$$p_r = g_r - \sum_{k=r+1}^{L_2(r)} u_{rk} p_k \tag{5-21}$$

The computing labor involved in the band algorithm is as follows:

Factorization: $\mathcal{O}(W^2 N) + \mathcal{O}(WN) + \ldots$

Forward and back solution: $\mathcal{O}(WN) + \mathcal{O}(N) + \ldots$

APPLICATION OF BAND ALGORITHM TO TWO-DIMENSIONAL PROBLEMS

Standard ordering

To cast eq. (5-5) with boundary conditions (2-17) into band matrix form, we proceed as follows. We choose an ordering of the points (i, j) so that:

$$r = i + I(j-1) \tag{5-22}$$

It will be convenient to define some coefficients:

$\alpha_i(k) = 0$ if $i = k$

$\alpha_i(k) = 1$ if $i \neq k$

Then the elements of **B** can be defined as follows:

$$b_{rr} = \alpha_i(1)AX_{i-\frac{1}{2},j} + \alpha_i(I)AX_{i+\frac{1}{2},j} + \alpha_j(1)AY_{i,j-\frac{1}{2}} + \alpha_j(J)AY_{i,j+\frac{1}{2}}$$

$$b_{r,r-1} = -\alpha_i(1)AX_{i-\frac{1}{2},j}$$

$$b_{r,r+1} = -\alpha_i(I)AX_{i+\frac{1}{2},j}$$

$$b_{r,r-I} = -\alpha_j(1)AY_{i,j-\frac{1}{2}}$$

$$b_{r,r+I} = -\alpha_j(J)AY_{i,j+\frac{1}{2}}$$

$$b_{rs} = 0, \quad \text{for } s \neq r-1, r, r+1, r-I, \text{ or } r+I$$

$$q_r = Q_{ij}$$

The total number of equations is $N = I \cdot J$. The width is $W = 2I + 1$. Hence the computing labor will be:

$$\mathcal{O}(I^2 \cdot IJ) + \mathcal{O}(I \cdot IJ) + \ldots = \mathcal{O}(I^3 J) + \mathcal{O}(I^2 J) + \ldots$$

Where $I \neq J$, clearly we should orient the rectangle so the shorter side is on the x-axis, in order to effect a very substantial reduction in computing time.

Nonstandard orderings

Price and Coats (1974) have shown that ordering the points of the computing rectangle in some sequence other than that given by eq. (5-22) can lead to very substantial savings in computing time. For some "diagonal" orderings, they have reduced the time by a factor of four for some two-dimensional problems, and by as much as a factor of six for some three-dimensional problems. George (1973) has proposed a "nested dissection" scheme for ordering the points which lead to theoretical estimates of computing time such as $\mathcal{O}(I^3)$ when $I = J$. These nonstandard orderings show considerable promise, therefore, in extending the applicability of direct solution algorithms to the solution of reservoir problems. They do increase the programming complexity of the algorithms very greatly, however, particularly for very large problems where the entire coefficient matrix cannot be kept in random-access memory. Hence, the potential savings in computer time obtainable through the use of these new orderings is uncertain at the present time.

ITERATIVE METHODS FOR SOLVING ELLIPTIC PROBLEMS

As we shall see in Chapter 6, a reservoir simulation consists of repeatedly solving large sets of difference equations, once per time step. Some of these equations are elliptic in character. For many large simulations involving many grid points, direct methods are too expensive to be practical, so iterative methods must be used. In the remainder of this chapter, we shall discuss three types of iteration for the solution of elliptic problems: (1) relaxation,

(2) alternating-direction, and (3) S.I.P. (strongly implicit procedure). All three types are in use throughout the industry, and any given simulator may use one or more of these types.

All iterative methods involve making some initial guess for the dependent variable or variables. This initial guess is usually the set of values obtained on the previous time step. Some method is then used to improve the guess with each iteration until, after a sufficient number of iterations, the simultaneous equations are satisfied to within some criterion of accuracy.

The superscript, k, is used to indicate the iteration count. Thus P_{ij}^k is the value of P_{ij} after k iterations. Each iteration method is described by an algorithm which transforms the vector, $\mathbf{p}^k$, whose elements are P_{ij}^k, into the vector, $\mathbf{p}^{k+1}$, whose elements are P_{ij}^{k+1}. The initial guess corresponds to $k=0$ and is therefore denoted by the initial vector, $\mathbf{p}^0$, with elements P_{ij}^0.

All iteration methods involve an amount of computing directly proportional to the number of grid points, $I \cdot J$, multiplied by the number of iterations.

Finally, all iteration methods involve one or more *iteration parameters*. These are multipliers of additional terms added to the iteration equations to accelerate the convergence. We shall discuss some of the factors involved in choosing these iteration parameters for each of the three methods.

POINT RELAXATION METHODS

Introduction

Equation (5-5) can be rewritten in the form:

$$d_{ij}P_{ij} - AX_{i-\frac{1}{2},j}P_{i-1,j} - AX_{i+\frac{1}{2},j}P_{i+1,j} - AY_{i,j-\frac{1}{2}}P_{i,j-1}$$
$$- AY_{i,j+\frac{1}{2}}P_{i,j+1} - Q_{ij} = 0 \qquad (5\text{-}23)$$

On the x-y plane, we can imagine that the point $(i+1, j)$ lies to the east of point (i, j). Similarly, the points $(i, j+1)$, $(i-1, j)$, and $(i, j-1)$ lie, respectively, to the north, west, and south of the center point (i, j). (See Fig. 6 or Fig. 8 in Chapter 2.) In this spirit, we can consider $AX_{i+\frac{1}{2},j}$ to be the "east" coefficient AE_{ij}, $AX_{i-\frac{1}{2},j}$ to be the "west" coefficient AW_{ij}, etc., and write eq. (5-23) in the following, more general, form:

$$d_{ij}P_{ij} - AW_{ij}P_{i-1,j} - AE_{ij}P_{i+1,j} - AS_{ij}P_{i,j-1} - AN_{ij}P_{i,j+1} - Q_{ij} = 0 \quad (5\text{-}24)$$

We may no longer require symmetry of the coefficient matrix, but we will require diagonal dominance, namely:

$$d_{ij} \geqslant AW_{ij} + AE_{ij} + AS_{ij} + AN_{ij}$$

Southwell relaxation

For a considerable period before the advent of electronic computers, many elliptic problems were being solved by hand. Gauss was probably the first to suggest the relaxation method; Seidel also suggested it. Southwell probably did the most to popularize the method and to foster its application to engineering problems.

Basically, the method consists in writing eq. (5-24) in residual form:

$$R_{ij} = d_{ij}P_{ij} - AW_{ij}P_{i-1,j} - AE_{ij}P_{i+1,j} - AS_{ij}P_{i,j-1} - AN_{ij}P_{i,j+1} - Q_{ij} = 0 \quad (5\text{-}25)$$

where R_{ij} is the *residual*. If P_{ij} is the solution to eq. (5-24), then each residual is, of course, zero. In general, however, R_{ij} is not zero, so one starts with the arbitrary initial guess for P_{ij} and evaluates R_{ij} for each point (i, j). He then scans the entire list of residuals for the one with the largest magnitude and, at that point (i, j), "relaxes" R_{ij} to zero by solving for P_{ij}:

$$P_{ij} = (AW_{ij}P_{i-1,j} + AE_{ij}P_{i+1,j} + AS_{ij}P_{i,j-1} + AN_{ij}P_{i,j+1} + Q_{ij})/d_{ij} \quad (5\text{-}26)$$

Obviously, changing P at the point (i, j) changes the residual for each of its neighbors. Hence, the R's need to be updated for each of the four surrounding points. Then the list of residuals is again scanned for the one with the largest magnitude and the process is repeated.

It was found by experience that the relaxation process can be speeded up by *overrelaxing*. That is, after finding the change in P_{ij} that would reduce the residual at that point to zero, one deliberately goes further and makes an even greater change in P_{ij}. For example, if P_{ij}^{old} is the old value, and we have just solved for P_{ij}^* by eq. (5-26), then the new value of P_{ij} is obtained by:

$$P_{ij}^{\text{new}} = P_{ij}^{\text{old}} + \beta(P_{ij}^* - P_{ij}^{\text{old}})$$

The quantity, β, is known as the overrelaxation parameter. If $\beta = 1$, we have normal relaxation, corresponding to use of eq. (5-26); if $\beta > 1$, we have overrelaxation; if $\beta < 1$, we have underrelaxation. Of course, if $\beta \neq 1$, then a new residual, not equal to zero, must be computed for the point (i, j) as well as for its four neighbors.

Obviously, the arithmetic involved in relaxation is quite elementary for a simple elliptic problem such as the solution of Laplace's equation with Dirichlet boundary conditions (i.e., where the function is known on the boundary). The residual equation (5-25) then simplifies to:

$$R_{ij} = 4P_{ij} - P_{i-1,j} - P_{i+1,j} - P_{i,j-1} - P_{i,j+1} = 0$$

It is easy to see that a change in P_{ij} of $+1$ will increase R_{ij} by 4 and will decrease the residual for each neighbor by 1. Thus it is very easy to keep track of all the residuals as well as to make use of overrelaxation.

Skill and judgment can contribute much to the efficiency of relaxation by hand. Various special techniques were evolved for speeding up convergence, such as block relaxation, where a group of points have their values of P simultaneously increased or decreased by the same amount.

However, the unsystematic nature of Southwell relaxation makes it ill-suited for use on automatic computers. The search for the largest residual is a relatively inefficient process; further, the judgment required for deciding when and how much to overrelax or underrelax, or when to use block relaxation, has not been reduced to rules that can be programmed. With the increased availability of high-speed computers, there seems to be little incentive for further study of this type of relaxation. Rather, the emphasis is on what might be called systematic relaxation, or iteration, methods.

Gauss-Seidel method (method of successive displacements)

The Gauss-Seidel method is the same as ordinary relaxation, except that the points are taken in an arbitrary, but fixed, order. As in Fig. 15, we shall consider the sequence wherein the points are taken in each row in order of increasing i, and then the rows are taken in order of increasing j. After traversing all the points, we start over again at the first point. A complete iteration consists of one trip through all the points. Recall that the superscript, k, refers to the number of complete iterations. Then we have:

$$P_{ij}^{k+1} = (AW_{ij}P_{i-1,j}^{k+1} + AE_{ij}P_{i+1,j}^{k} + AS_{ij}P_{i,j-1}^{k+1} + AN_{ij}P_{i,j+1}^{k} + Q_{ij})/d_{ij} \quad (5\text{-}27)$$

Note that the result would be the same if we took the points in order of increasing j in each column, and then took the columns in order of increasing i. Thus we may characterize the ordering of points as being from "southwest to northeast".

For the solution of general sets of simultaneous equations, this method is referred to as the Gauss-Seidel method. When applied to elliptic difference equations, it may also be referred to as the Liebman method.

Successive overrelaxation (SOR)

This is similar to Southwell overrelaxation, but with the points taken in systematic order. Let β be the relaxation parameter. Then we have:

$$P_{ij}^{k+1} = P_{ij}^{k} + \beta[\{(AW_{ij}P_{i-1,j}^{k+1} + AE_{ij}P_{i+1,j}^{k} + AS_{ij}P_{i,j-1}^{k+1} + AN_{ij}P_{i,j+1}^{k} + Q_{ij})/d_{ij}\} - P_{ij}^{k}] \quad (5\text{-}28)$$

This method has also been referred to as the extrapolated Liebman method. For $\beta = 1$, it reduces to the Gauss-Seidel method.

Method of simultaneous displacements (Jacobi method)

In this method one solves for P_{ij} by eq. (5-26) but does not use the improved values until *after* a complete iteration. Thus the iteration equation is:

$$P_{ij}^{k+1} = (AW_{ij}P_{i-1,j}^k + AE_{ij}P_{i+1,j}^k + AS_{ij}P_{i,j-1}^k + AN_{ij}P_{i,j+1}^k + Q_{ij})/d_{ij} \quad (5\text{-}29)$$

Note that the sequence in which the points are taken is immaterial in this method.

Matrix representation of point-iteration methods

Let **p** be a vector whose elements are P_{ij}, and **q** be a vector whose elements are Q_{ij}. Then the difference equation to be solved, eq. (5-24), may be represented in matrix form by:

Ap = **q** (5-30)

where **A** is a coefficient matrix, similar in structure to that shown in Fig. 17. The coefficient matrix can be split into three matrices:

A = **L** + **D** + **U** (5-31)

where **L** is a strictly lower triangular matrix containing the elements of **A** below the main diagonal, namely $-AW_{ij}$ and $-AS_{ij}$; **D** is a diagonal matrix whose elements are the diagonal elements of **A**, namely d_{ij}; **U** is a strictly upper triangular matrix containing the elements of **A** above the main diagonal, namely, $-AE_{ij}$ and $-AN_{ij}$.

Jacobi method

If we multiply each eq. (5-29) by d_{ij}, then the Jacobi method may be represented in matrix form by:

Dp^{k+1} = $-$ **Lp**^{k} $-$ **Up**^{k} + **q**

or, alternatively, by:

Dp^{k+1} = **Dp**^{k} $-$ **Ap**^{k} + **q** (5-32)

Let **p**$^\infty$ be the solution after convergence. Then, since **p**$^{k+1}$ = **p**^{k} at convergence, eq. (5-32) becomes:

$-$ **Ap**$^\infty$ + **q** = 0 (5-33)

Comparison with eq. (5-30), whose solution we desire, shows that **p**$^\infty$ (and therefore the elements, P_{ij}^∞) satisfies the original difference equation (5-24). Since the solution to (5-24) is unique, it follows that if the iteration converges, it converges to the solution.

Define an error at each point by:

$$E_{ij}^k = P_{ij}^k - P_{ij}^\infty \quad (5\text{-}34)$$

and the corresponding error vector by:

$$e^k = p^k - p^\infty \tag{5-35}$$

Substitution of eqs. (5-33) and (5-35) into eq. (5-32) gives the following error equation:

$$De^{k+1} = De^k - Ae^k \tag{5-36}$$

The error equation can be made more concise by utilizing the concepts of the identity matrix and the inverse of a matrix. The identity matrix, denoted by **I**, is a diagonal matrix whose elements are all ones on the main diagonal and zero elsewhere. The inverse of a matrix **M** is denoted by M^{-1}, and is defined by:

$$M^{-1}M = MM^{-1} = I$$

It is easy to show that the inverse of a diagonal matrix, **D**, with elements d_{ij}, is simply another diagonal matrix containing the reciprocals of d_{ij} along the main diagonal.

By premultiplying eq. (5-36) by D^{-1}, we obtain:

$$e^{k+1} = e^k - D^{-1}Ae^k$$

or:

$$e^{k+1} = G_J e^k \tag{5-37}$$

where:

$$G_J = I - D^{-1}A \tag{5-38}$$

is the Jacobi iteration matrix. Note that this matrix operates on the error, not on the solution; the properties of this matrix determine how rapidly the error diminishes with each iteration and, therefore, how rapidly the iteration converges.

Successive overrelaxation

If we multiply each eq. (5-28) by d_{ij}, the successive overrelaxation method is represented in matrix form by:

$$Dp^{k+1} = Dp^k + \beta(-Lp^{k+1} - Up^k + q - Dp^k) \tag{5-39}$$

By substituting eqs. (5-31), (5-33), and (5-35) into eq. (5-39), we obtain the following error equation:

$$De^{k+1} = De^k - \beta[Le^{k+1} + (U+D)e^k] \tag{5-40}$$

This may be rearranged to give:

$$(\beta L + D)e^{k+1} = -[(\beta - 1)D + \beta U]e^k \tag{5-41}$$

Premultiplying by the inverse of $(\beta L + D)$ gives:

$$e^{k+1} = G_{SOR} e^k \tag{5-42}$$

where:
$$G_{SOR} = -(\beta L + D)^{-1}[(\beta - 1)D + \beta U] \tag{5-43}$$
is the SOR iteration matrix.

Convergence rate for Jacobi iteration by harmonic analysis

Of the various point-relaxation methods described above, the Jacobi method is the easiest to analyze for rate of convergence. Let us consider a simple case, namely the solution of eq. (5-5) with the AX and AY constant (though not necessarily the same). This corresponds to a uniformly anisotropic medium with constant Δx and Δy, but with the possibility that $\Delta x \neq \Delta y$. Since $AW_{ij} = AE_{ij} = AX$, $AS_{ij} = AN_{ij} = AY$, and $d_{ij} = 2(AX + AY)$, then the (matrix) error equation (5-36) may be written in the difference form:

$$2(AX + AY)E_{ij}^{k+1} = 2(AX + AY)E_{ij}^{k} + (AX)\delta_x^2 E_{ij}^k + (AY)\delta_y^2 E_{ij}^k \tag{5-44}$$

where δ_x^2 and δ_y^2 are second-difference operators such that:

$$\delta_x^2 P_{ij} = P_{i-1,j} - 2P_{ij} + P_{i+1,j} \tag{5-45}$$
$$\delta_y^2 P_{ij} = P_{i,j-1} - 2P_{ij} + P_{i,j+1} \tag{5-46}$$

Because of the assumption of constant AX and AY, we can use harmonic analysis to analyze the convergence, in a manner similar to a von Neumann stability analysis. However, we need to be more precise now in choosing functions that satisfy the boundary conditions, since the boundary conditions affect the rate of convergence. Rather than use exponential functions, we use sine functions when Dirichlet boundary conditions are specified, i.e., when the solution is specified on the boundary. (This corresponds to zero error there.) For Neumann boundary conditions (such as zero normal derivatives on the boundary), cosine functions are used.

We assume that the error is a sum of components, chosen according to the type of boundary condition used.

Error expansion for Dirichlet boundary conditions, point-centered grid

$$E_{ij}^k = \sum_{p,q} (E_{pq}^k)_{ij} = \sum_{p=1}^{I-1} \sum_{q=1}^{J-1} \gamma_{pq}^k \sin(p\pi i/I) \sin(q\pi j/J) \tag{5-47}$$

Error expansion for Neumann boundary conditions, point-centered grid

To satisfy the discrete Neumann boundary condtions of eq. (2-18), we would use:

$$E_{ij}^k = \sum_{p=0}^{I} \sum_{q=0}^{J} \gamma_{pq}^k a_i^* b_j^* \cos(p\pi i/I) \cos(q\pi j/J)$$

where $a_i^* = \frac{1}{2}$ for $i = 0$ or I; $a_i^* = 1$ for $1 \leq i \leq I-1$; $b_j^* = \frac{1}{2}$ for $j = 0$ or J; $b_j^* = 1$ for $1 \leq j \leq J-1$.

Error expansion for Neumann boundary conditions, block-centered grid

To satisfy the discrete Neumann boundary conditions of eq. (2-17), we use:

$$E_{ij}^k = \sum_{p=0}^{I-1} \sum_{q=0}^{J-1} \gamma_{pq}^k \cos(p\pi i'/I) \cos(q\pi j'/J) \tag{5-48}$$

where $i' = i - \frac{1}{2}$ and $j' = j - \frac{1}{2}$.

Convergence analysis for Neumann boundary conditions

Since Neumann boundary conditions are of greater significance than Dirichlet boundary conditions in reservoir problems, we devote most of our attention to them. There is a trivial difference in convergence rate for the two types of discrete Neumann boundary conditions that correspond to eqs. (2-17) and (2-18). The analysis is somewhat simpler for the block-centered grid, so we shall confine our attention to the error expansion given in eq. (5-48). Now:

$$\delta_x^2[\cos(p\pi i'/I)] = \cos[(p\pi i'/I) - (p\pi/I)] - 2\cos(p\pi i'/I) + \cos[(p\pi i'/I) + (p\pi/I)]$$

$$\delta_x^2[\cos(p\pi i'/I)] = \cos(p\pi i'/I)[2\cos(p\pi/I) - 2]$$

$$\delta_x^2[\cos(p\pi i'/I)] = -4\cos(p\pi i'/I)\sin^2(p\pi/2I) \tag{5-49}$$

Similarly:

$$\delta_y^2[\cos(q\pi j'/J)] = -4\cos(q\pi j'/J)\sin^2(q\pi/2J) \tag{5-50}$$

Substitution into eq. (5-44) and elimination of common factors gives, for each component of (5-48):

$$2(AX + AY)(\gamma_{pq}^{k+1} - \gamma_{pq}^k) = -4\gamma_{pq}^k[AX\sin^2(p\pi/2I) + AY\sin^2(q\pi/2J)]$$

which reduces to:

$$\gamma_{pq}^{k+1}/\gamma_{pq}^k = 1 - \frac{2}{AX + AY}[AX\sin^2(p\pi/2I) + AY\sin^2(q\pi/2J)] \tag{5-51}$$

For convergence, it is necessary that each component of the error be reduced in magnitude. Thus we require:

$$|\gamma_{pq}^{k+1}/\gamma_{pq}^k| < 1$$

Young (1962) defines the rate of convergence to be:

$$R = -\max_{p,q} \ln|\gamma_{pq}^{k+1}/\gamma_{pq}^k| \tag{5-52}$$

Hence we must look at the worst case (i.e., the component with the largest ratio). This will occur when both sines of eq. (5-51) are close to zero (i.e., $p = 0, q = 0; p = 1, q = 0; p = 0, q = 1; p = 1, q = 1$) or when both sines are close to unity (i.e., $p = I-1, q = J-1$).

Let us examine the various combinations of p and q. The case of $p = 0$, $q = 0$ gives:

$$\gamma_{0,0}^{k+1}/\gamma_{0,0}^{k} = 1$$

This component corresponds to the arbitrary constant of integration for an elliptic problem with Neumann boundary conditions, so the fact that it cannot be reduced is of no consequence.

For the case of $p = 1, q = 0$, we have:

$$\sin^2 (p\pi/2I) = \sin^2 (\pi/2I) \approx \pi^2/4I^2$$

$$\sin^2 (q\pi/2J) = 0$$

and:

$$\gamma_{pq}^{k+1}/\gamma_{pq}^{k} \approx 1 - \frac{2}{AX + AY}(AX\pi^2/4I^2) = 1 - \frac{\pi^2}{2(AX + AY)}(AX/I^2) \quad (5\text{-}53)$$

For the case of $p = 0, q = 1$, we have:

$$\sin^2 (p\pi/2I) = 0$$

$$\sin^2 (q\pi/2J) = \sin^2 (\pi/2J) \approx \pi^2/4J^2$$

and:

$$\gamma_{pq}^{k+1}/\gamma_{pq}^{k} \approx 1 - \frac{\pi^2}{2(AX + AY)}(AY/J^2) \quad (5\text{-}54)$$

For the case of $p = 1, q = 1$, we have:

$$\sin^2 (p\pi/2I) = \sin^2 (\pi/2I) \approx \pi^2/4I^2$$

$$\sin^2 (q\pi/2J) \approx \pi^2/4J^2$$

and:

$$\gamma_{pq}^{k+1}/\gamma_{pq}^{k} \approx 1 - \frac{\pi^2}{2(AX + AY)}\left(\frac{AX}{I^2} + \frac{AY}{J^2}\right) \quad (5\text{-}55)$$

Finally, for the case of $p = I-1; q = J-1$, we have:

$$\sin^2 (p\pi/2I) = \sin^2 [(I-1)\pi/2I] = \cos^2 (\pi/2I) \approx 1 - (\pi^2/4I^2)$$

$$\sin^2 (q\pi/2J) \approx 1 - (\pi^2/4J^2)$$

and:

$$\gamma_{pq}^{k+1}/\gamma_{pq}^{k} \approx 1 - \frac{2AX}{AX + AY}\left(1 - \frac{\pi^2}{4I^2}\right) - \frac{2AY}{AX + AY}\left(1 - \frac{\pi^2}{4J^2}\right)$$

$$\gamma_{pg}^{k+1}/\gamma_{pg}^{k} \approx -1 + \frac{\pi^2}{2(AX+AY)}\left(\frac{AX}{I^2} + \frac{AY}{J^2}\right) \tag{5-56}$$

Both eq. (5-55) and (5-56) are of the same magnitude, though of opposite sign. However, both eq. (5-53) and (5-54) are larger (closer to unity) and therefore are controlling. We then have:

$$\max_{p,q} |\gamma_{pq}^{k+1}/\gamma_{pq}^{k}| \approx 1 - \frac{\pi^2}{2(AX+AY)} \cdot \min\,[(AX/I^2),(AY/J^2)] \tag{5-57}$$

Inasmuch as the number being subtracted from one on the right-hand side is small, the rate of convergence can be well approximated by:

$$R(G_J) \approx \frac{\pi^2}{2(AX+AY)} \cdot \min\,[(AX/I^2),(AY/J^2)] \tag{5-58}$$

Effect of anistropy on convergence rate for Neumann boundary conditions

For problems in a square grid (where $I = J$), eq. (5-58) reduces to:

$$R(G_J) \approx \frac{\pi^2/2I^2}{1 + \max\,[(AX/AY),(AY/AX)]} \tag{5-59}$$

It can be seen that the rate of convergence for Jacobi iteration decreases very significantly as the degree of anisotropy increases. This phenomenon is directly related to the fact that Neumann boundary conditions were assumed, as we shall see in the next section.

Convergence rate for Dirichlet boundary conditions

As pointed out previously, eq. (5-47) should be used for the error expansion when Dirichlet boundary conditions are specified. Expressions analogous to eqs. (5-49) and (5-50) hold for the sine functions; these are:

$$\delta_x^2[\sin\,(p\pi i/I)] = -4\sin\,(p\pi i/I)\sin^2\,(p\pi/2I)$$

$$\delta_y^2[\sin\,(q\pi j/J)] = -4\sin\,(q\pi j/J)\sin^2\,(q\pi/2J)$$

Hence eq. (5-51) is still valid for the ratio $\gamma_{pq}^{k+1}/\gamma_{pq}^{k}$. However, there are no components involving $p = 0$ or $q = 0$, so that eqs. (5-55) and (5-56) provide the maximum values for the magnitude of this ratio. Thus the rate of convergence can be approximated by:

$$R(G_J) \approx \frac{\pi^2}{2(AX+AY)}\left(\frac{AX}{I^2} + \frac{AY}{J^2}\right) \tag{5-60}$$

For problems involving a square grid ($I = J$), this reduces to the simple expression:

$$R(G_J) \approx \pi^2/2I^2 \tag{5-61}$$

In contrast to problems involving the Neumann boundary condition, we see no effect of anisotropy here.

For isotropic problems, an additional conclusion can be obtained. Comparison of eq. (5-61) with (5-59) shows that when $AX = AY$, the rate of convergence with Dirichlet boundary conditions is twice that achieved when Neumann boundary conditions are used.

Convergence rate for Jacobi iteration by eigenvalue analysis

While harmonic analysis provides a suitable method for analyzing the convergence of Jacobi iteration (at least for uniform AX and AY), it cannot be used for analyzing successive overrelaxation. For that purpose we will require matrix eigenvalue analysis. We introduce this type of analysis by applying it, by way of illustration, to the simpler Jacobi iteration, and then we shall apply it to SOR in the next section. The eigenvalue analysis will provide some additional insight into Jacobi iteration under conditions when the simplifying assumptions of uniform AX and AY do not hold.

Definition of eigenvalues and eigenvectors
In general, the matrix problem:

$$\mathbf{Mv} = \lambda \mathbf{v} \qquad (5\text{-}62)$$

has only the trivial solution, $\mathbf{v} = \mathbf{0}$. But for a finite number of values of λ, known as the eigenvalues of $\mathbf{M}$, there exist nontrivial solutions, $\mathbf{v}$, which are the eigenvectors of $\mathbf{M}$. Each eigenvector is associated with an eigenvalue.

Alternatively, eq. (5-62) may be written:

$$(\mathbf{M} - \lambda \mathbf{I})\mathbf{v} = \mathbf{0} \qquad (5\text{-}63)$$

Since $\mathbf{v}$ is nonzero when λ is an eigenvalue, it follows that the determinant of $(\mathbf{M} - \lambda \mathbf{I})$ must be zero. Thus, an alternate defining relationship for λ to be an eigenvalue of $\mathbf{M}$ is that it satisfies the equation:

$$\det (\mathbf{M} - \lambda \mathbf{I}) = 0 \qquad (5\text{-}64)$$

Relation between convergence and eigenvalues
The general iteration equation may be written in the form:

$$e^{k+1} = \mathbf{G} e^k \qquad (5\text{-}65)$$

The eigenvalue problem associated with the iteration matrix, $\mathbf{G}$, is:

$$\mathbf{G} \mathbf{v}_{pq} = \mu_{pq} \mathbf{v}_{pq} \qquad (5\text{-}66)$$

If we let p take on the values $0, 1, \ldots, I-1$, and q take on the values $0, 1, \ldots, J-1$, then there are $I \cdot J$ independent eigenvectors $\mathbf{v}_{pq}$. For the problems we are interested in, these eigenvectors form a complete set, so that any arbitrary vector can be expanded as a sum of these component eigenvectors. In particular, the error vectors may be expanded as:

$$\mathbf{e}^k = \sum_p \sum_q \gamma_{pq}^k \mathbf{v}_{pq}$$

and:

$$\mathbf{e}^{k+1} = \sum_p \sum_q \gamma_{pq}^{k+1} \mathbf{v}_{pq}$$

If we substitute these into eq. (5-65), we obtain:

$$\sum_p \sum_q \gamma_{pq}^{k+1} \mathbf{v}_{pq} = \sum_p \sum_q \gamma_{pq}^k \mathbf{G} \mathbf{v}_{pq}$$

Use of eq. (5-66) then gives:

$$\sum_p \sum_q \gamma_{pq}^{k+1} \mathbf{v}_{pq} = \sum_p \sum_q \gamma_{pq}^k \mu_{pq} \mathbf{v}_{pq}$$

so that, for each component, we have:

$$\gamma_{pq}^{k+1}/\gamma_{pq}^k = \mu_{pq} \tag{5-67}$$

We have thus established a direct correspondence between the convergence ratio for a particular component and the corresponding eigenvalue. Young's definition for convergence rate, eq. (5-52), may now be written as:

$$R(\mathbf{G}) = -\max_{p,q} \ln |\mu_{pq}| \tag{5-68}$$

where μ_{pq} is an eigenvalue of the iteration matrix, $\mathbf{G}$.

Application to Jacobi iteration
Equation (5-38) gives the Jacobi iteration matrix:

$$\mathbf{G}_J = \mathbf{I} - \mathbf{D}^{-1}\mathbf{A}$$

If we let μ be an eigenvalue of $\mathbf{G}_J$, then by eq. (5-64) we have:

$$\det(\mathbf{I} - \mathbf{D}^{-1}\mathbf{A} - \mu \mathbf{I}) = 0$$

Since this can be rewritten:

$$\det[\mathbf{D}^{-1}\mathbf{A} - (1-\mu)\mathbf{I}] = 0$$

it follows that $(1 - \mu)$ is an eigenvalue of $\mathbf{D}^{-1}\mathbf{A}$.

One may characterize $\mathbf{D}^{-1}\mathbf{A}$ as the "problem" matrix, since it is not associated with any particular iteration method. If we designate the eigenvalue of the problem matrix by $\lambda(\mathbf{D}^{-1}\mathbf{A})$, we have:

$$\lambda = 1 - \mu$$

or:

$$\mu(G_J) = 1 - \lambda(D^{-1}A) \tag{5-69}$$

Thus, given the eigenvalues of the problem matrix, one can obtain the eigenvalues of the Jacobi iteration matrix.

The eigenvalues, λ, correspond to nontrivial solutions of the matrix problem:

$$D^{-1}Av = \lambda v$$

which can also be written as:

$$Av = \lambda Dv \tag{5-70}$$

While these eigenvalues may be somewhat difficult to obtain for the general coefficient matrix, A, they are easy to evaluate for the special case where AX and AY are uniform and $d_{ij} = 2(AX + AY)$. If we consider the vector v_{pq} to have components V_{ij}^{pq}, then eq. (5-70) can be written in the difference form:

$$-(AX)\delta_x^2 V_{ij}^{pq} - (AY)\delta_x^2 V_{ij}^{pq} = 2(AX + AY)\lambda_{pq} V_{ij}^{pq} \tag{5-71}$$

It happens that the components of eq. (5-48) are eigenfunctions of $D^{-1}A$, so we can write:

$$V_{ij}^{pq} = \cos(p\pi i'/I) \cos(q\pi j'/J) \tag{5-72}$$

Substituting eq. (5-72) into (5-71), making use of eqs. (5-49) and (5-50), and cancelling common factors gives:

$$4AX \sin^2(p\pi/2I) + 4AY \sin^2(q\pi/2J) = 2(AX + AY)\lambda_{pq}$$

or:

$$\lambda_{pq}(D^{-1}A) = 2[AX \sin^2(p\pi/2I) + AY \sin^2(q\pi/2J)]/[AX + AY] \tag{5-73}$$

Substitution of eq. (5-73) into (5-69) then gives:

$$\mu_{pq}(G_J) = 1 - 2[AX \sin^2(p\pi/2I) + AY \sin^2(q\pi/2J)]/[AX + AY]$$

Since $\mu_{pq} = \gamma_{pq}^{k+1}/\gamma_{pq}^k$, this agrees with our earlier result for the convergence ratio, i.e., eq. (5-51).

Thus, for this special case of Jacobi iteration with uniform AX and AY, we have shown that the eigenvalue method for analyzing convergence rate gives the same result as that obtained by harmonic analysis.

Convergence rate for successive overrelaxation

We now have most of the machinery required for determining the rate of convergence for successive overrelaxation. We do need one more piece of equipment, however; this is a particular property of some matrices, known as Property A.

Property A

Property A, invented by Young, states that the matrix $A = L + D + U$ can be transformed, by reordering its columns and rows, into a "diagonally block-tridiagonal form", as shown in Fig. 20 (Forsythe and Wasow, 1960, p. 243). The D_r's are diagonal submatrices of various orders, made up of the elements of D; the L_r's are *rectangular* submatrices made up of the elements of L; the U_r's are rectangular submatrices made up of the elements of U. Note that the columns and rows of A must be reordered in exactly the same way. The reordering is not unique, so that more than one reordering of the columns and rows of A will put it into the form of Fig. 20. Since reordering of columns and rows of a matrix does not alter its determinant, we have the important result that the reordered matrix has the same eigenvalues as the original matrix, A.

$$\begin{bmatrix} D_1 & U_1 & & & & & \\ L_2 & D_2 & U_2 & & & & \\ & L_3 & D_3 & U_3 & & & \\ & & & \cdot & & & \\ & & & & \cdot & & \\ & & & & & L_s & D_s \end{bmatrix}$$

Fig. 20. Diagonally block-tridiagonal form of matrix.

For example, consider the twelve-point problem shown in Fig. 15. Take the points in the order 1, 2, 5, 3, 6, 9, 4, 7, 10, 8, 11, 12. This ordering is obtained by starting at the southwest corner and taking points on successive diagonals. The reordered matrix, shown in Fig. 21, is indeed seen to be in the diagonally block-tridiagonal form of Fig. 20.

Eigenvalues of SOR iteration matrix

Let ω be an eigenvalue of G_{SOR}. Then, by eq. (5-64):

$$\det \{G_{SOR} - \omega I\} = 0 \qquad (5\text{-}74)$$

Substitution for G_{SOR}, given by eq. (5-43), gives:

$$\det \{-(\beta L + D)^{-1}[(\beta - 1)D + \beta U] - \omega I\} = 0$$

Since the determinant of the product of matrices is the product of the determinants, and since $\det (\beta L + D)^{-1} \neq 0$, it follows that:

$$\det [(\beta - 1)D + \beta U + \omega(\beta L + D)] = 0$$

or:

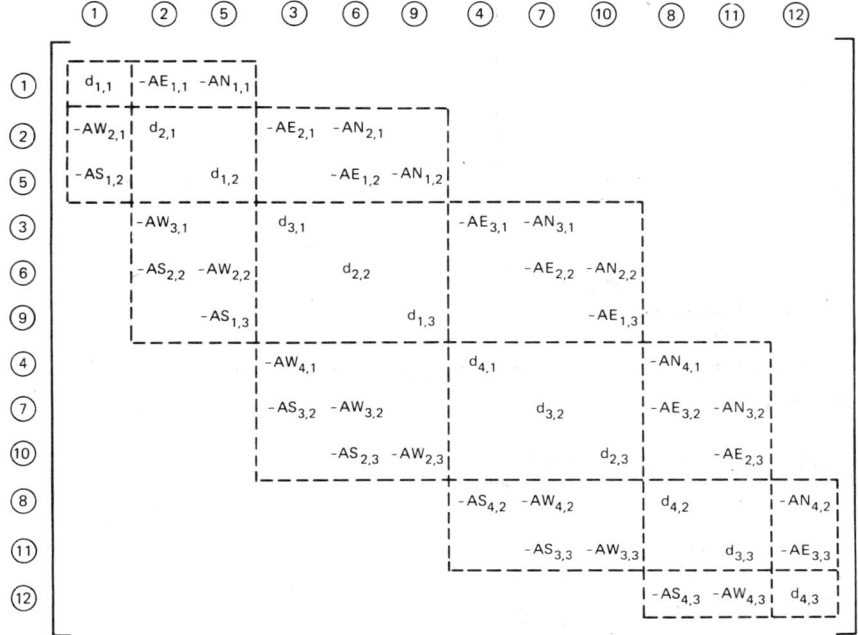

Fig. 21. Reordered coefficient matrix for $I = 4$, $J = 3$.

$$\det(\omega L + aD + U) = 0 \tag{5-75}$$

where:

$$a = (\beta + \omega - 1)/\beta \tag{5-76}$$

Now, because of Property A, the matrix of eq. (5-75) can be written in the diagonally block-tridiagonal form shown in Fig. 22. Let Γ be the diagonal matrix shown in Fig. 23. (Note that for each r, I_r has the same order as the diagonal submatrix, D_r, of Fig. 20.) Carry out the multiplication,

$$\begin{bmatrix} aD_1 & U_1 & & & & \\ \omega L_2 & aD_2 & U_2 & & & \\ & \omega L_3 & aD_3 & U_3 & & \\ & & & \bullet & & \\ & & & & \bullet & \\ & & & & \omega L_s & aD_s \end{bmatrix}$$

Fig. 22. Matrix $(\omega L + aD + U)$ written in diagonally block-tridiagonal form.

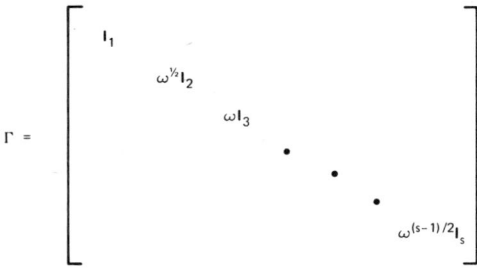

Fig. 23. Definition of diagonal matrix, Γ.

$\Gamma^{-1}(\omega L + aD + U)\Gamma$; the result is shown in Fig. 24. We can see, from inspection of Fig. 24, that:

$$\Gamma^{-1}(\omega L + aD + U)\Gamma = \omega^{\frac{1}{2}}L + aD + \omega^{\frac{1}{2}}U$$

Then:

$\det(\omega^{\frac{1}{2}}L + aD + \omega^{\frac{1}{2}}U) = 0$

from which it follows that:

$\det[\omega^{\frac{1}{2}}(L + D + U) + (a - \omega^{\frac{1}{2}})D] = 0$

$\det[D^{-1}A + (a\omega^{-\frac{1}{2}} - 1)I] = 0$

or:

$\det(I - D^{-1}A - a\omega^{-\frac{1}{2}}I) = 0$

From eq. (5-38), we know that $I - D^{-1}A$ is G_J, the Jacobi iteration matrix, the eigenvalues of which we have already designated by μ. Thus:

$a\omega^{-\frac{1}{2}} = \mu(G_J)$

With the substitution of the definition for a, eq. (5-76), we then have:

$$\beta + \omega - 1 = \mu\beta\omega^{\frac{1}{2}} \qquad (5\text{-}77)$$

$$\begin{bmatrix} aD_1 & \omega^{\frac{1}{2}}U_1 & & & & \\ \omega^{\frac{1}{2}}L_2 & aD_2 & \omega^{\frac{1}{2}}U_2 & & & \\ & \omega^{\frac{1}{2}}L_3 & aD_3 & \omega^{\frac{1}{2}}U_3 & & \\ & & & \bullet & & \\ & & & & \bullet & \\ & & & & \omega^{\frac{1}{2}}L_s & aD_s \end{bmatrix}$$

Fig. 24. Product of $\Gamma^{-1}(\omega L + aD + U)\Gamma$, written in diagonally block-tridiagonal form.

or:
$$\omega^{\frac{1}{2}} = \tfrac{1}{2}\beta\mu \pm \tfrac{1}{2}[\beta^2\mu^2 - 4(\beta-1)]^{\frac{1}{2}} \tag{5-78}$$

More on eigenvalues of Jacobi iteration matrix

Using a weaker form of Property A, we can show that if μ is an eigenvalue of G_J, then so is $-\mu$. Let us use a "checkerboard" ordering of points. The points are imagined to be colored alternately red and black, and we take all the red points in sequence followed by all the black points in sequence. (In the twelve-point case of Fig. 15, this could be the order 1, 3, 6, 8, 9, 11, 2, 4, 5, 7, 10, 12.) Then the matrix $A = L + D + U$ can be written in the form:

$$A = \begin{bmatrix} D_1 & F \\ E & D_2 \end{bmatrix}$$

where D_1 and D_2 are diagonal submatrices containing elements from D, and E and F are (possibly) rectangular submatrices containing only elements from L and U.

Let μ be an eigenvalue of $G_J = I - D^{-1}A$. Then:

$$\det(D - A - \mu D) = 0$$

or:

$$\det(-L - U - \mu D) = 0 \tag{5-79}$$

If the rows and columns are reordered, using the checkerboard ordering, then the matrix of eq. (5-79) can be written:

$$-L - U - \mu D = \begin{bmatrix} -\mu D_1 & -F \\ -E & -\mu D_2 \end{bmatrix}$$

Let I_1 be an identity matrix with the same order as D_1, and I_2 be an identity matrix with the same order as D_2. Define Γ to be the diagonal matrix:

$$\Gamma = \begin{bmatrix} I_1 & 0 \\ 0 & -I_2 \end{bmatrix}$$

Then, by carrying out the indicated multiplications, we can obtain the product:

$$\Gamma^{-1}(-L - U - \mu D)\Gamma = \begin{bmatrix} -\mu D_1 & F \\ E & -\mu D_2 \end{bmatrix}$$

$$= L + U - \mu D$$

$$= A - D - \mu D$$

from which it follows that:

$$\det(A - D - \mu D) = \det[\Gamma^{-1}(-L - U - \mu D)\Gamma]$$
$$= \det(-L - U - \mu D) = 0$$

or:

$$\det(I - D^{-1}A + \mu I) = 0$$

Therefore, $-\mu$ is also an eigenvalue of G_J.

Relation between convergence of SOR and convergence of Jacobi method

Suppose that the Jacobi method does not converge. This means that some $|\mu| > 1$. We ask whether successive overrelaxation will converge. Since μ and $-\mu$ both occur as eigenvalues, we take the positive one, $\mu > 1$, and take the positive root in the expression for ω, eq. (5-78):

$$\max \omega^{\frac{1}{2}} = \tfrac{1}{2}|\beta\mu| + \tfrac{1}{2}[\beta^2\mu^2 - 4(\beta - 1)]^{\frac{1}{2}}$$

From the assumption that $|\mu| > 1$, it follows that:

$$\max \omega^{\frac{1}{2}} > \tfrac{1}{2}|\beta| + \tfrac{1}{2}[\beta^2 - 4\beta + 4]^{\frac{1}{2}}$$

$$\max \omega^{\frac{1}{2}} > \tfrac{1}{2}|\beta| + \tfrac{1}{2}|\beta - 2|$$

We need to examine three cases:

(a) $\beta \leqslant 0$. Then $|\beta| = -\beta$, $|\beta - 2| = 2 - \beta$, and
$$\max \omega^{\frac{1}{2}} > \tfrac{1}{2}(2 - 2\beta) = 1 - \beta.$$

(b) $0 \leqslant \beta \leqslant 2$. Then $|\beta| = \beta$, $|\beta - 2| = 2 - \beta$, and
$$\max \omega^{\frac{1}{2}} > \tfrac{1}{2}(\beta + 2 - \beta) = 1.$$

(c) $2 \leqslant \beta$. Then $|\beta| = \beta$, $|\beta - 2| = \beta - 2$, and
$$\max \omega^{\frac{1}{2}} > \tfrac{1}{2}(2\beta - 2) = \beta - 1.$$

In each case:

$$\max \omega^{\frac{1}{2}} \geqslant 1$$

Thus, successive overrelaxation can *never* be a convergent iteration (for real β) when the Jacobi method itself diverges. The advantage of successive overrelaxation lies in accelerating the rate of convergence of an already convergent process. From now on, we consider only $|\mu| < 1$.

Convergence rate of method of successive displacements

Consider first, the method of successive displacements, i.e., SOR with $\beta = 1$. Then, from eq. (5-78), we have:

$$\omega^{\frac{1}{2}} = \tfrac{1}{2}\mu + \tfrac{1}{2}(\mu^2)^{\frac{1}{2}}$$

$$\max \omega^{\frac{1}{2}} = \max |\mu|$$

$$\max |\omega| = \max \mu^2$$

Since, by eq. (5-68):

$$\begin{aligned} R(G_{SD}) &= -\ln \max |\omega| \\ &= -2 \ln \max |\mu| \\ &= 2R(G_J) \end{aligned} \quad (5\text{-}80)$$

The method of successive displacements converges exactly twice as fast as the Jacobi method.

Optimum parameter for SOR

What is the optimum value of β to use for successive overrelaxation? To guide the search, eq. (5-78) is plotted in Fig. 25, which shows $|\omega|$ vs. β for different values of μ. First, we see the previous result that $|\omega| > 1$ for $\mu > 1$.

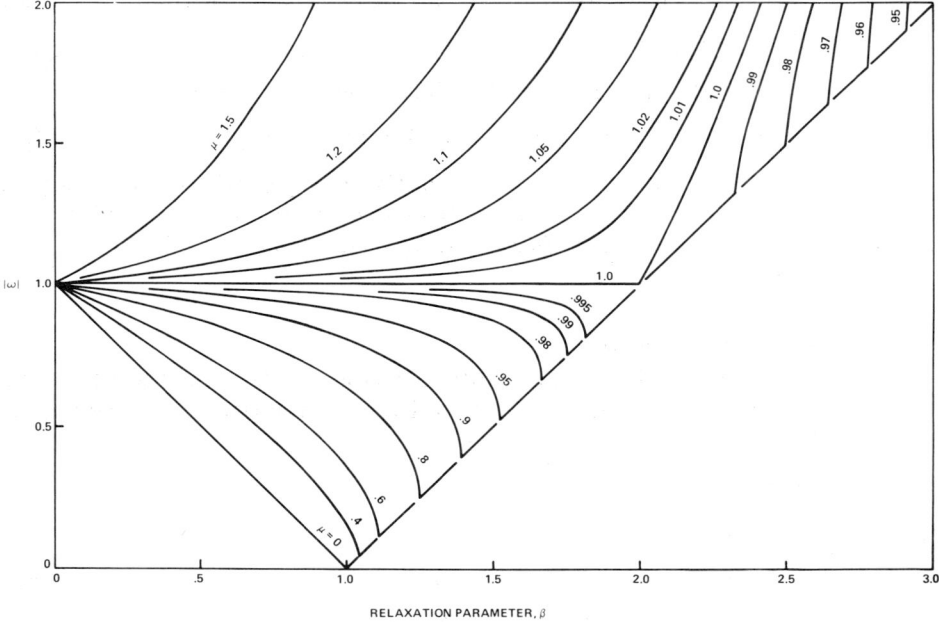

Fig. 25. Solution of eq. (5-78), showing variation of $|\omega|$ with β for various values of μ. ω is eigenvalue of G_{SOR}; μ is eigenvalue of G_J; β is SOR relaxation parameter.

We also see that $|\omega| = \mu^2$ at $\beta = 1$. We note that for $|\mu| < 1$, $|\omega|$ is always greater than μ^2 for $\beta < 1$, and is greater than 1 for $\beta > 2$. It is clear that the optimum β will lie between 1 and 2.

It turns out that the break in each curve, which is where $|\omega|$ is minimized, occurs at that value of β for which the discriminant, $\beta^2 \mu^2 - 4(\beta - 1)$, is zero.

Thus:

$$\mu^2 \beta_{opt}^2 - 4\beta_{opt} + 4 = 0 \tag{5-81}$$

or:

$$\beta_{opt} = 2[1 \pm (1-\mu^2)^{\frac{1}{2}}]/\mu^2$$

The negative root should be taken, since $\beta_{opt} < 2$. Then:

$$\beta_{opt} = 2[1 - (1-\mu^2)^{\frac{1}{2}}]/\mu^2 \tag{5-82}$$

From eq. (5-81):

$$\mu^2 \beta_{opt} = 4(\beta_{opt} - 1)/\beta_{opt} = 4(1 - \beta_{opt}^{-1})$$

Combination with eq. (5-82) yields the alternative expression:

$$\beta_{opt} = 2/[1 + (1-\mu^2)^{\frac{1}{2}}] \tag{5-83}$$

Since the discriminant is zero at this value of β, we obtain from eq. (5-78):

$$|\omega| = \beta_{opt}^2 \mu^2/4 = \beta_{opt}(\beta_{opt}\mu^2)/4$$

which, combined with eqs. (5-82) and (5-83), yields the result:

$$\max |\omega_{opt}| = [1 - (1 - \max \mu^2)^{\frac{1}{2}}]/[1 + (1 - \max \mu^2)^{\frac{1}{2}}] \tag{5-84}$$

The straight-line portion of the curves in Fig. 25 corresponds to the region in which the discriminant is negative. It can be shown for this region that $|\omega|$ is independent of μ and satisfies the simple equation

$$|\omega| = \beta - 1$$

Comparison of convergence rates

How much better SOR is than Jacobi iteration and the method of successive displacements can be seen as follows. In most practical cases, we expect $\max |\mu|$ to be slightly less than one. Let

$$\max |\mu| = 1 - \epsilon$$

For the Jacobi method:

$$R(G_J) = -\max \ln |\mu| = -\ln (1-\epsilon) \approx \epsilon \tag{5-85}$$

For successive displacements (SOR with $\beta = 1$):

$$R(G_{SD}) = 2R(G_J) \approx 2\epsilon$$

For optimum successive overrelaxation:

$$\max \mu^2 = (1-\epsilon)^2 \approx 1 - 2\epsilon$$

$$(1 - \max \mu^2)^{\frac{1}{2}} \approx (2\epsilon)^{\frac{1}{2}}$$

From eq. (5-84):

$$\max |\omega_{opt}| \approx [1-(2\epsilon)^{\frac{1}{2}}]/[1+(2\epsilon)^{\frac{1}{2}}] \approx 1-2(2\epsilon)^{\frac{1}{2}}$$

and:

$$R(G_{SOR})_{opt} = -\max \ln |\omega_{opt}| \approx 2^{3/2}\epsilon^{\frac{1}{2}} \tag{5-86}$$

Substitution of eq. (5-85) into (5-86) gives:

$$R(G_{SOR})_{opt} \approx 2^{3/2}[R(G_J)]^{\frac{1}{2}} \tag{5-87}$$

For small ϵ, this is a very large improvement. To illustrate this, we consider the solution of Laplace's equation in a square with Neumann boundary conditions. Consider three mesh sizes, $I = J = 10$, $I = J = 32$, and $I = J = 100$. For this simple example, $\Delta x = \Delta y$, and $AX = AY$. Then, by eqs. (5-57) and (5-67):

$$\max |\mu| = 1 - (\pi^2/4)(1/I^2)$$

The results are summarized in Table V. The last line of the table shows the potential improvement in convergence rate that SOR possesses over Jacobi iteration.

TABLE V

Comparison of point iteration methods

Mesh size, $I \times J$	10 × 10	32 × 32	100 × 100
Jacobi iteration			
$\max |\mu|$	0.9753	0.99759	0.999753
$\epsilon = \pi^2/4I^2$	0.0247	0.00241	0.000247
$R(G_J) = -\max \ln |\mu|$	0.0250	0.00241	0.000247
Method of successive displacement			
$R(G_{SD}) = 2R(G_J)$	0.0500	0.00482	0.000494
Optimum successive overrelaxation			
β_{opt}, from eq. (5-83)	1.638	1.870	1.957
$\max |\omega_{opt}|$, from eq. (5-84)	0.6383	0.8702	0.9565
$R(G_{SOR})_{opt} = -\ln \max |\omega_{opt}|$	0.4489	0.1390	0.04443
$2^{3/2}\epsilon^{\frac{1}{2}} \approx R(G_{SOR})_{opt}$	0.4443	0.1388	0.04443
Improvement $= R(G_{SOR})_{opt}/R(G_J)$	17.97	57.61	180.1

The advantage of selecting the optimum β when using SOR is obvious. Unfortunately, as can be seen from Fig. 25, it depends very critically on having a good estimate for $\max |\mu|$, and this is hard to come by for problems more complex than those involving uniform AX and AY in a rectangle. One rule to follow in estimating β_{opt} is to try to be on the high side, as can be seen from the curves for $|\omega|$ vs. β. Some discussions of the problem of choosing β_{opt} are given by Young (1962, 1971) and Forsythe and Wasow (1960).

Effects of anisotropy and boundary conditions on convergence rate of SOR

Inasmuch as the rate of convergence of SOR is directly related to the rate of convergence of Jacobi iteration, we can immediately assess the effects of anisotropy and boundary conditions on SOR.

For isotropic problems in a square grid, the above conclusion that Jacobi iteration for problems with Dirichlet boundary conditions converges twice as fast as for problems with Neumann boundary conditions leads to the conclusion, via eq. (5-87), that optimal SOR will also converge faster, but only by a factor of $2^{\frac{1}{2}}$. For problems with Dirichlet boundary conditions, anisotropy should not significantly affect the rate of convergence of SOR. On the other hand, just as for Jacobi iteration, we can expect that anisotropy will significantly slow down the convergence of SOR when Neumann boundary conditions are used.

LINE RELAXATION METHODS

Introduction

The previously described point-iteration methods are also called explicit methods because, in each step of the process, one unknown, P_{ij}^{k+1}, is solved for explicitly. An implicit method is one in which several unknowns at adjacent points are solved for simultaneously. These are referred to as block iterative methods. The most important example of this is where all the points on a line are solved for simultaneously. We can again distinguish two classes of processes, simultaneous displacements and successive displacements.

Line simultaneous displacements (line-Jacobi)

In the present discussion we take our lines to be columns (i.e., constant x). For simultaneous displacements by columns, all the unknowns on the ith column, $P_{i,1}, P_{i,2}, \ldots, P_{i,J}$, are solved for implicitly. Thus, instead of eq. (5-29), the iteration equation is:

$$P_{ij}^{k+1} = (AW_{ij}P_{i-1,j}^{k} + AE_{ij}P_{i+1,j}^{k} + AS_{ij}P_{i,j-1}^{k+1} + AN_{ij}P_{i,j+1}^{k+1} + Q_{ij})/d_{ij} \quad (5\text{-}88)$$

In this method, the order in which the columns are solved is immaterial.

For each column, the equations form a tridiagonal system. We have already considered the solution of such equations by the tridiagonal algorithm in Chapter 3. Using that algorithm, the computing labor per line is directly proportional to the number of points on the line, so that the work required for one line-Jacobi iteration should be only slightly greater than that required for one point-Jacobi iteration.

Line successive overrelaxation (LSOR)

Analogous to point successive relaxation, we consider solving the columns in sequence, and use the newest values of the unknowns in the equations for the next column. Then the LSOR equation (without overrelaxation) would be:

$$P^*_{ij} = (AW_{ij}P^{k+1}_{i-1,j} + AE_{ij}P^k_{i+1,j} + AS_{ij}P^*_{i,j-1} + AN_{ij}P^*_{i,j+1} + Q_{ij})/d_{ij} \quad (5\text{-}89)$$

Again, the extrapolation equation is:

$$P^{k+1}_{ij} = P^k_{ij} + \beta(P^*_{ij} - P^k_{ij}) \quad (5\text{-}90)$$

but this is not applied until after eq. (5-89) is solved for the entire line.

Convergence rate of line-Jacobi iteration

With Neumann boundary conditions

The potential improvement of line iteration over point iteration can perhaps best be illustrated by harmonic analysis of the convergence rate for the line-Jacobi method. Again we assume the simplified case that AX and AY are uniform, and that $d_{ij} = 2(AX + AY)$. Then the error equation corresponding to eq. (5-88) can be written:

$$2(AX)E^{k+1}_{ij} = 2(AX)E^k_{ij} + (AX)\delta^2_x E^k_{ij} + (AY)\delta^2_y E^{k+1}_{ij}$$

Again, we assume that the error is a sum of components, where each component is a product of cosines, as in eq. (5-48). Then, upon substitution of eqs. (5-48), (5-49) and (5-50), we have for each component:

$$2AX(\gamma^{k+1}_{pq} - \gamma^k_{pq}) = -4AX\gamma^k_{pq}\sin^2(p\pi/2I) - 4AY\gamma^{k+1}_{pq}\sin^2(q\pi/2J)$$

Thus:

$$\mu_{pq}(G_{LJ}) = \gamma^{k+1}_{pq}/\gamma^k_{pq} = \frac{1 - 2\sin^2(p\pi/2I)}{1 + 2(AY/AX)\sin^2(q\pi/2J)} \quad (5\text{-}91)$$

We shall see later that the case of $AX \gg AY$ is not of interest, since line iteration by rows rather than by columns would then be used. In searching for the maximum eigenvalue, μ, we consider again the same combinations of p and q as we examined for point-Jacobi iteration. In each case, I and J are assumed large.

For the various combinations, eq. (5-91) reduces to

$$(p = 1, q = 0): \mu_{1,0} = 1 - 2\sin^2(\pi/2I) \approx 1 - (\pi^2/2I^2) \quad (5\text{-}92)$$

$$(p=0, q=1): \mu_{0,1} = \frac{1}{1 + 2(AY/AX)\sin^2(\pi/2J)}$$

$$\approx \frac{1}{1 + (AY/AX)(\pi^2/2J^2)} \tag{5-93}$$

$$(p=1, q=1): \mu_{1,1} = \frac{1 - 2\sin^2(\pi/2I)}{1 + 2(AY/AX)\sin^2(\pi/2J)}$$

$$\approx \frac{1 - (\pi^2/2I^2)}{1 + (AY/AX)(\pi^2/2J^2)} \tag{5-94}$$

$$(p=I-1, q=J-1): \mu_{I-1,J-1} = \frac{1 - 2\cos^2(\pi/2I)}{1 + 2(AY/AX)\cos^2(\pi/2J)}$$

$$\approx \frac{-1 + (\pi^2/2I^2)}{1 + (AY/AX)[2 - (\pi^2/2J^2)]} \tag{5-95}$$

It follows from the assumptions of $AY \geqslant AX$ and $J \gg 1$ that the denominator of eq. (5-95) is larger than the denominator of (5-94). Hence $\mu_{I-1,J-1}$ is less in magnitude than $\mu_{1,1}$, which in turn is less than $\mu_{0,1}$. Thus either $\mu_{0,1}$ or $\mu_{1,0}$ is the maximum eigenvalue, and the rate of convergence is, using eq. (5-68), given by:

$$R(\mathbf{G}_{\text{LJ}}) \approx \min\left[(\pi^2/2I^2); \; (AY/AX)(\pi^2/2J^2)\right] \tag{5-96}$$

For isotropic problems in a square grid ($AX = AY$; $I = J$), the convergence rate is:

$$R(\mathbf{G}_{\text{LJ}}) \approx \pi^2/2I^2 \tag{5-97}$$

which shows, by comparison with eq. (5-59), that line-Jacobi is twice as fast as point-Jacobi when Neumann boundary conditions are used.

For highly anisotropic problems ($AY \gg AX$), $\mu_{0,1}$ will be considerably smaller than $\mu_{1,0}$, so that the convergence rate will *still* be given by eq. (5-97). Comparison with eq. (5-59) shows this to be a considerable improvement over point-Jacobi, since the convergence rate for line-Jacobi is thus essentially independent of the degree of anisotropy, while that for point-Jacobi decreases very significantly.

We have yet to dispose of the anisotropic case where $AY \ll AX$. For this case, eq. (5-96) reduces to:

$$R(\mathbf{G}_{\text{LJ}}) \approx (AY/AX)(\pi^2/2J^2)$$

indicating a much slower rate of convergence than that given by eq. (5-97). For this type of anisotropy, it is clear that the use of columns for the direction of the line iteration is the wrong choice; rather, the points should be solved simultaneously along the rows.

With Dirichlet boundary conditions

As pointed out in the discussion of point-Jacobi iteration, components with $p = 0$ or $q = 0$ are not involved when Dirichlet boundary conditions are specified. Then $\mu_{1,1}$ is the largest eigenvalue in magnitude, and the rate of convergence is given by:

$$R(\mathbf{G_{LJ}}) \approx (\pi^2/2I^2) + (AY/AX)(\pi^2/2J^2) \qquad (5\text{-}98)$$

For isotropic problems in a square grid, this reduces to:

$$R(\mathbf{G_{LJ}}) \approx \pi^2/I^2$$

which, by comparison with eq. (5-61), shows again that line-Jacobi is twice as fast as point-Jacobi.

For anisotropic problems, however, eq. (5-98) shows a marked effect of the degree of anisotropy. For large AY/AX, the rate is approximately proportional to AY/AX; this contrasts with point-Jacobi, for which we found essentially no effect of the degree of anisotropy when Dirichlet boundary conditions are used.

Acceleration of convergence with Neumann boundary conditions

1-D method of additive corrections

We have seen in the previous section that, for anisotropic problems with Neumann boundary conditions, line iteration is considerably faster than point iteration, but not as fast as would be the case if Dirichlet boundary conditions were specified. Watts (1971, 1973) proposed using the method of additive corrections to accelerate the convergence of LSOR for anisotropic problems with Neumann boundary conditions. This method comes close to achieving the convergence rate obtainable with Dirichlet boundary conditions.

The combination of LSOR with additive corrections is designated as LSORC. Before we consider LSORC, let us first discuss the method of additive corrections and examine its effect on line-Jacobi iteration. The combination of LJ with additive corrections is designated LJC.

Watts's method of additive corrections consists in adding to each P in a column a constant, α_i, so that a "corrected" value of $P_{ij}^{c,k+1}$ is used as the starting point for the next iteration. Thus:

$$P_{ij}^{c,k+1} = P_{ij}^{k+1} + \alpha_i \qquad (5\text{-}99)$$

The α's are chosen so that the *sum* of the residuals for each column is reduced to zero. To obtain the equations for the α's, we write the following two equations for residuals similar to eq. (5-25):

$$R_{ij}^{k+1} = d_{ij}P_{ij}^{k+1} - AW_{ij}P_{i-1,j}^{k+1} - AE_{ij}P_{i+1,j}^{k+1} - AS_{ij}P_{i,j-1}^{k+1} - AN_{ij}P_{i,j+1}^{k+1} - Q_{ij}$$

$$R_{ij}^{c,k+1} = d_{ij}P_{ij}^{c,k+1} - AW_{ij}P_{i-1,j}^{c,k+1} - AE_{ij}P_{i+1,j}^{c,k+1} - AS_{ij}P_{i,j-1}^{c,k+1} - AN_{ij}P_{i,j+1}^{c,k+1}$$
$$- Q_{ij} \qquad (5\text{-}100)$$

Combining with eq. (5-99) gives:

$$R_{ij}^{c,k+1} = R_{ij}^{k+1} - AW_{ij}\alpha_{i-1} + (d_{ij} - AS_{ij} - AN_{ij})\alpha_i - AE_{ij}\alpha_{i+1}$$

Then the condition that the sum of residuals in a column be reduced to zero, i.e.:

$$\sum_{j=1}^{J} R_{ij}^{c,k+1} = 0 \qquad (5\text{-}101)$$

leads to the system of equations:

$$\alpha_{i-1} \sum_{j=1}^{J} AW_{ij} + \alpha_i \sum_{j=1}^{J} (AS_{ij} + AN_{ij} - d_{ij}) + \alpha_{i+1} \sum_{j=1}^{J} AE_{ij} = \sum_{j=1}^{J} R_{ij}^{k+1}$$

which can be easily solved for the α's by the tridiagonal algorithm given in Chapter 3.

To understand why this method of additive corrections accelerates the convergence of line-Jacobi iteration (and also LSOR), we begin by expressing the residuals in terms of the error. Substitute eq. (5-34) into (5-100) and use the fact that P_{ij}^{∞} satisfies eq. (5-24) to obtain:

$$R_{ij}^{c,k+1} = d_{ij}E_{ij}^{c,k+1} - AW_{ij}E_{i-1,j}^{c,k+1} - AE_{ij}E_{i+1,j}^{c,k+1} - AS_{ij}E_{i,j-1}^{c,k+1} - AN_{ij}E_{i,j+1}^{c,k+1}$$

The assumption that AX and AY are uniform and that $d_{ij} = 2(AX + AY)$ yields:

$$R_{ij}^{c,k+1} = -(AX)\delta_x^2 E_{ij}^{c,k+1} - (AY)\delta_y^2 E_{ij}^{c,k+1}$$

Then, using the error expansion of eq. (5-48) and substituting eqs. (5-49) and (5-50), we have:

$$R_{ij}^{c,k+1} = 4 \sum_{p=0}^{I-1} \sum_{q=0}^{J-1} \gamma_{pq}^{c,k+1} \cos(p\pi i'/I) \cos(q\pi j'/J)[AX \sin^2(p\pi/2I)$$
$$+ AY \sin^2(q\pi/2J)]$$

The condition that the sum of residuals along a column be zero, eq. (5-101), yields:

$$\sum_{j=1}^{J} \sum_{p=0}^{I-1} \sum_{q=0}^{J-1} \gamma_{pq}^{c,k+1} \cos(p\pi i'/I) \cos(q\pi j'/J)[AX \sin^2(p\pi/2I)$$
$$+ AY \sin^2(q\pi/2J)] = 0$$

Reordering the summations:

$$\sum_{p=0}^{I-1} \sum_{q=0}^{J-1} \gamma_{pq}^{c,k+1} \cos(p\pi i'/I)[AX \sin^2(p\pi/2I) + AY \sin^2(q\pi/2J)]$$

$$\cdot \sum_{j=1}^{J} \cos(q\pi j'/J) = 0$$

But, for $q \neq 0$, it can be shown that:

$$\sum_{j=1}^{J} \cos(q\pi\{j - \tfrac{1}{2}\}/J) = 0$$

For $q = 0$ the sum is, of course, equal to J. Then:

$$\sum_{p=0}^{I-1} \gamma_{p,0}^{c,k+1} \cos(p\pi i'/I)[AX \sin^2(p\pi/2I)]J = 0$$

from which it follows that:

$$\gamma_{p,0}^{c,k+1} = 0 \qquad (5\text{-}102)$$

In other words, the method of additive corrections annihilates all the components for which $q = 0$.

As a consequence of this, the eigenvalues associated with the $q = 0$ components do not need to be included in the search for the maximum eigenvalue. The maximum eigenvalue of line-Jacobi with correction is thus given by eq. (5-93), and the rate of convergence is given (for large J) by:

$$R(G_{\text{LJC}}) \approx (AY/AX)(\pi^2/2J^2) \qquad (5\text{-}103)$$

For large AY/AX, we see that this is almost as large as the rate for line-Jacobi iteration for Dirichlet problems, given by eq. (5-98).

For isotropic problems with Neumann boundary conditions, comparison with line-Jacobi iteration, whose rate is given by eq. (5-96), shows essentially no difference. We can conclude that the method of additive corrections provides the most improvement in convergence rate for anisotropic problems, while not causing any change in rate for isotropic problems.

Since there are no components involving $q = 0$ associated with problems using Dirichlet boundary conditions, there is no incentive for applying the method of additive corrections to such problems.

The above analysis, showing that the method of additive corrections annihilates the $q = 0$ components, is applicable only for Jacobi iteration, and cannot be directly extended to LSOR. The components (i.e., eigenvectors) for LSOR are not the same as the eigenvectors for LJ. However, Watts (1973) has shown that the $q = 0$ eigenvectors for LSOR are also annihilated by the method of additive corrections, so that the maximum eigenvalue for LJC *can* be used to calculate the maximum eigenvalue for LSORC, as we shall do in the next section.

It would appear sufficient to apply the additive correction only once, say prior to the first iteration. In theory, once the eigenvectors for $q = 0$ are annihilated, they stay annihilated. In practice, as a result of round-off, some of these eigenvectors are reintroduced into the solution during the iteration, so that it is necessary to perform the additive correction step every several iterations (Watts suggests every ten iterations). Further, the theory is strictly applicable only for the case of uniform AX and AY, and experiments with nonuniformly anisotropic problems show that it is necessary to do the additive correction step frequently, in some cases as often as every iteration.

2-D method of additive corrections

It is clear that the one-dimensional additive corrections method described above will have maximum benefit when the anisotropy is completely in one direction, with $AY \gg AX$. In many real reservoir situations such as one-well coning problems, where the radial grid spacing is very small near the well and very large away from the well, the direction of higher transmissibility may vary from one part of the computing region to another. Thus we may have $AY > AX$ in one subregion and $AX > AY$ in another. For such situations, Settari and Aziz (1973) proposed a two-dimensional additive corrections procedure. In its simplest form, their technique can be described as two one-dimensional corrections in sequence, one of which reduces the sum of residuals to zero in one direction (say columns, as described above) followed by another step which reduces the sum of residuals to zero in the other direction. For the line-Jacobi iteration, this removes all the components involving both $p = 0$ and $q = 0$. For LSOR, Watts has shown that one of the sets of eigenvectors is annihilated (either $p = 0$ or $q = 0$); the other set is greatly reduced but not completely annihilated (J.W. Watts, personal communication, 1976).

Settari and Aziz (1973) have also suggested that the methods of additive corrections have greater generality, in that they can be applied to other iteration schemes besides line relaxation methods.

Convergence rates of LSOR and LSORC

Methods similar to those used for point-SOR can be used to analyze line-SOR, both without and with additive corrections. Results similar to eqs. (5-83) and (5-84) are obtained, namely:

$$\beta(\mathbf{G}_{\text{LSOR}})_{\text{opt}} = \frac{2}{1 + [1 - \max \mu^2(\mathbf{G}_{\text{LJ}})]^{\frac{1}{2}}} \tag{5-104}$$

and:

$$\max |\omega(\mathbf{G}_{\text{LSOR}})_{\text{opt}}| = \frac{1 - [1 - \max \mu^2(\mathbf{G}_{\text{LJ}})]^{\frac{1}{2}}}{1 + [1 - \max \mu^2(\mathbf{G}_{\text{LJ}})]^{\frac{1}{2}}} \tag{5-105}$$

Further, a result similar to eq. (5-87) is obtained:

$$R(G_{LSOR})_{opt} \approx 2^{3/2}[R(G_{LJ})]^{\frac{1}{2}} \tag{5-106}$$

Similar equations hold for LSORC. In particular:

$$R(G_{LSORC})_{opt} \approx 2^{3/2}[R(G_{LJC})]^{\frac{1}{2}} \tag{5-107}$$

By substituting the various asymptotic rates of convergence already derived for line-Jacobi iterations into eqs. (5-106) and (5-107), the following asymptotic rates of convergence for LSOR and LSORC are easily obtained.

With Neumann boundary conditions

$$R(G_{LSOR})_{opt} \approx (2\pi/I) \cdot \min\,[1;(AY/AX)^{\frac{1}{2}}(I/J)] \tag{5-108}$$

$$R(G_{LSORC})_{opt} \approx (2\pi/I)(AY/AX)^{\frac{1}{2}}(I/J) \tag{5-109}$$

With Dirichlet boundary conditions

$$R(G_{LSOR})_{opt} = R(G_{LSORC})_{opt} \approx (2\pi/I)[1+(AY/AX)(I^2/J^2)]^{\frac{1}{2}} \tag{5-110}$$

Summary of convergence rates for point and line relaxation methods

To facilitate the comparison of the various relaxation methods presented in this chapter, the asymptotic convergence rates are listed in Table VI. To simplify the comparisons, a square grid ($I = J$) is assumed.

TABLE VI

Asymptotic convergence rates, $R(G)$, for various relaxation methods. Square grid ($I = J$), and $AY \geq AX$, are assumed

Method	Boundary conditions	
	Neumann	Dirichlet
Point-Jacobi (J)	$(\pi^2/2I^2)[1+(AY/AX)]^{-1}$	$\pi^2/2I^2$
Line-Jacobi (LJ)	$\pi^2/2I^2$	$(\pi^2/2I^2)[1+(AY/AX)]$
Line-Jacobi, corrected (LJC)	$(\pi^2/2I^2)(AY/AX)$	$(\pi^2/2I^2)[1+(AY/AX)]$
Optimal SOR	$(2\pi/I)[1+(AY/AX)]^{-\frac{1}{2}}$	$2\pi/I$
Optimal LSOR	$2\pi/I$	$(2\pi/I)[1+(AY/AX)]^{\frac{1}{2}}$
Optimal LSORC	$(2\pi/I)(AY/AX)^{\frac{1}{2}}$	$(2\pi/I)[1+(AY/AX)]^{\frac{1}{2}}$

Examination of this table reconfirms some of the general conclusions already reached. For isotropic problems, Jacobi iterations with Dirichlet boundary conditions converge twice as fast as Jacobi iterations with Neumann boundary conditions, while each SOR method with Dirichlet boundary conditions converges $2^{1/2}$ times as fast as the same SOR method with Neumann boundary conditions. Again for isotropic problems, line-

Jacobi iteration converges twice as fast as point-Jacobi iteration, while LSOR converges $2^{1/2}$ times as fast as SOR. Finally, for highly anisotropic problems with Neumann boundary conditions, the table clearly shows the improvements in convergence rates brought about by the use of additive corrections.

ALTERNATING-DIRECTION ITERATION (A.D.I.)

Formulation of Peaceman-Rachford iteration procedure

As discussed in Chapter 3, the Peaceman-Rachford alternating-direction *implicit* procedure was first proposed for the solution of unsteady-state, or parabolic, problems. Since the solution of an elliptic problem is the steady-state (or infinite-time) solution of a corresponding parabolic problem, it was soon realized that an alternating-direction approach could be used for solving elliptic problems, wherein each time step corresponds to one step of an iteration. The size of the time step is arbitrary in the approach to the steady-state solution. Hence, time-step size can be considered to be an iteration parameter, chosen to accelerate the convergence of the iteration to the steady-state solution.

The alternating-direction *iteration* procedure of Peaceman and Rachford (1955), as applied to the solution of eq. (5-5), is now given. It is analogous to eq. (3-54). Recall that the superscript k indicates the iteration number. The iteration proceeds in two steps:

(1) the x-sweep, implicit in the x-direction, wherein the equations for *each row* of points (i.e., in the x-direction) form one tridiagonal system of equations which is easily solved by the tridiagonal algorithm presented in Chapter 3;

(2) the y-sweep, implicit in the y-direction, wherein the equations for *each column* of points (i.e., in the y-direction) also form a tridiagonal system of equations which is easily solved.

x-sweep:

$$AX_{i+\frac{1}{2},j}(P^{k+1}_{i+1,j} - P^{k+1}_{ij}) - AX_{i-\frac{1}{2},j}(P^{k+1}_{ij} - P^{k+1}_{i-1,j}) + AY_{i,j+\frac{1}{2}}(P^k_{i,j+1} - P^k_{ij})$$
$$- AY_{i,j-\frac{1}{2}}(P^k_{ij} - P^k_{i,j-1}) + Q_{ij} = S_{ij}\beta_k(P^{k+1}_{ij} - P^k_{ij}) \qquad (5\text{-}111)$$

y-sweep:

$$AX_{i+\frac{1}{2},j}(P^{k+1}_{i+1,j} - P^{k+1}_{ij}) - AX_{i-\frac{1}{2},j}(P^{k+1}_{ij} - P^{k+1}_{i-1,j}) + AY_{i,j+\frac{1}{2}}(P^{k+2}_{i,j+1} - P^{k+2}_{ij})$$
$$- AY_{i,j-\frac{1}{2}}(P^{k+2}_{ij} - P^{k+2}_{i,j-1}) + Q_{ij} = S_{ij}\beta_k(P^{k+2}_{ij} - P^{k+1}_{ij}) \qquad (5\text{-}112)$$

β_k, the iteration parameter, corresponds to the reciprocal of the pseudo-time-step size. The subscript, k, on β indicates the possibility of changing that parameter for each double sweep. S_{ij} is a "normalizing factor" defined by:

$$S_{ij} = AX_{i-\frac{1}{2},j} + AX_{i+\frac{1}{2},j} + AY_{i,j-\frac{1}{2}} + AY_{i,j+\frac{1}{2}} \qquad (5\text{-}113)$$

The necessity for such a normalizing factor can be seen from the situation where AX and AY are equal and constant. In that case, we should like the result of applying eqs. (5-111) and (5-112) to depend only on β_k and to be independent of the constant chosen for AX and AY.

The matrix formulation of eqs. (5-111) and (5-112) is obtained by splitting the coefficient matrix into a "horizontal" matrix H and a "vertical" matrix **V**:

$$\mathbf{A} = \mathbf{H} + \mathbf{V} \tag{5-114}$$

where **H** contains only terms involving AX, and **V** contains only terms involving AY. In addition, let us also define the diagonal matrix **D** whose elements consist of the terms S_{ij}. Then the matrix formulation of eqs. (5-111) and (5-112) is:

$$\mathbf{H}\mathbf{p}^{k+1} + \mathbf{V}\mathbf{p}^k + \mathbf{q} = \beta_k \mathbf{D}(\mathbf{p}^{k+1} - \mathbf{p}^k) \tag{5-115}$$

$$\mathbf{H}\mathbf{p}^{k+1} + \mathbf{V}\mathbf{p}^{k+2} + \mathbf{q} = \beta_k \mathbf{D}(\mathbf{p}^{k+2} - \mathbf{p}^{k+1}) \tag{5-116}$$

While alternating-direction iteration has been found very useful for solving elliptic problems, even under rather general conditions, mathematical analysis of the method has been somewhat limited and requires various simplifying assumptions. For example, if the matrices **H** and **V** commute, i.e., if:

$$\mathbf{HV} = \mathbf{VH}$$

it can be demonstrated that the procedure will converge for any sequence of parameters, β_k. The most common situation, where AX and AY are not constant but vary with x and y, leads to noncommutivity of **H** and **V**, and it is indeed possible to choose sequences of β_k that will cause the alternating-direction iteration to diverge. It has been shown, however, for *any* matrices **H** and **V** (commutative or not), that alternating-direction iteration will converge for any *single* parameter β. The convergence will be extremely slow, however, and it has been found, both by experiment and by analysis, that a judicious choice of a sequence of parameters β_k will greatly accelerate the convergence.

A more complete discussion of alternating-direction iteration may be found in Varga (1962) or Young (1971).

Convergence analysis

Choice of parameters

We can analyze the convergence of A.D.I. by harmonic analysis, provided we assume that AX and AY are constant (though not necessarily the same). Then eqs. (5-111) and (5-112) may be written:

$$(AX)\delta_x^2 P_{ij}^{k+1} + (AY)\delta_y^2 P_{ij}^k + Q_{ij} = S\beta_k(P_{ij}^{k+1} - P_{ij}^k) \tag{5-117}$$

$$(AX)\delta_x^2 P_{ij}^{k+1} + (AY)\delta_y^2 P_{ij}^{k+2} + Q_{ij} = S\beta_k(P_{ij}^{k+2} - P_{ij}^{k+1}) \tag{5-118}$$

where $S = 2(AX + AY)$, and δ_x^2 and δ_y^2 are the second-difference operators defined by eqs. (5-45) and (5-46).

Let P_{ij}^∞ be the solution after convergence. Then:

$$(AX)\delta_x^2 P_{ij}^\infty + (AY)\delta_y^2 P_{ij}^\infty + Q_{ij} = 0 \tag{5-119}$$

The error at each point is defined by eq. (5-34). Substitution of eqs. (5-34) and (5-119) into eqs. (5-117) and (5-118) gives the following error equations:

$$(AX)\delta_x^2 E_{ij}^{k+1} + (AY)\delta_y^2 E_{ij}^k = S\beta_k(E_{ij}^{k+1} - E_{ij}^k) \tag{5-120}$$

$$(AX)\delta_x^2 E_{ij}^{k+1} + (AY)\delta_y^2 E_{ij}^{k+2} = S\beta_k(E_{ij}^{k+2} - E_{ij}^{k+1}) \tag{5-121}$$

We proceed with the harmonic analysis to examine the convergence in a manner similar to that which we used for the Jacobi iteration methods. We shall assume Neumann boundary conditions, and use the error expansion of eq. (5-48), where each component is a product of cosines. Then, upon substitution of eqs. (5-48), (5-49), and (5-50) into eqs. (5-120) and (5-121), we have for each component:

$$-4AX\gamma_{pq}^{k+1}\sin^2(p\pi/2I) - 4AY\gamma_{pq}^k\sin^2(q\pi/2J) = S\beta_k(\gamma_{pq}^{k+1} - \gamma_{pq}^k)$$

$$-4AX\gamma_{pq}^{k+1}\sin^2(p\pi/2I) - 4AY\gamma_{pq}^{k+2}\sin^2(q\pi/2J) = S\beta_k(\gamma_{pq}^{k+2} - \gamma_{pq}^{k+1})$$

These equations can be solved, respectively, for the ratios $\gamma_{pq}^{k+1}/\gamma_{pq}^k$ and $\gamma_{pq}^{k+2}/\gamma_{pq}^{k+1}$, which can then be multiplied together to give:

$$\frac{\gamma_{pq}^{k+2}}{\gamma_{pq}^k} = \frac{\beta_k - X_p}{\beta_k + X_p} \cdot \frac{\beta_k - Y_q}{\beta_k + Y_q} \tag{5-122}$$

where:

$$X_p = \frac{2AX}{AX + AY}\sin^2(p\pi/2I) \tag{5-123}$$

$$Y_q = \frac{2AY}{AX + AY}\sin^2(q\pi/2J) \tag{5-124}$$

Now p can take on the values $0, 1, 2, \ldots, I-1$, and q can take on the values $0, 1, 2, \ldots, J-1$. Ideally, we could try to pick β_k in such a way that some components will vanish for each k. For example, all the components $E_{1,q}$ (i.e., $p = 1$) will vanish for:

$$\beta_k = X_1 = \frac{2AX}{AX + AY}\sin^2(\pi/2I)$$

Let us therefore consider a sequence of parameters:

$$\beta_k = X_p = \frac{2AX}{AX+AY} \sin^2(p\pi/2I), \quad p = 1, 2, 3, \ldots, I-1 \qquad (5\text{-}125)$$

(Note that for strictly elliptic problems arising from incompressible flow problems, we cannot choose $\beta_k = 0$, as that will cause division by zero at the end of each line in the tridiagonal algorithm.) This sequence of parameters would annihilate all components for which $p \geqslant 1$ and $q \geqslant 0$. Components for which $p = 0$ would not be annihilated; for this we could choose another sequence of parameters:

$$\beta_k = Y_q = \frac{2AY}{AX+AY} \sin^2(q\pi/2J), \quad q = 1, 2, 3, \ldots, J-1 \qquad (5\text{-}126)$$

This would annihilate all components for which $q \geqslant 1$ and $p \geqslant 0$. This leaves only the component for which $p = q = 0$. We see from eqs. (5-122)–(5-124) that for any choice of $\beta_k > 0$:

$$\gamma_{0,0}^{k+2}/\gamma_{0,0}^{k} = 1$$

As already pointed out in the discussion of point-Jacobi iteration, the $p = 0$, $q = 0$ component corresponds to the arbitrary constant of integration for an elliptic problem with Neumann boundary conditions, and the fact that it cannot be reduced or annihilated is of no consequence.

As a practical matter, we do not actually attempt to annihilate specific components. High-frequency components which might be annihilated at one stage of the iteration would reappear due to round-off error. Furthermore, the above analysis is highly idealized because of the restrictive nature of the assumption of uniform AX and AY. Consequently, we merely use eqs. (5-125) and (5-126) as indicative of the "spectrum" that β_k should cover. We assume that I and J are sufficiently large that approximations may be used for the sines. Thus we have that β_k should span the two intervals:

$$m_1 = \frac{2AX}{AX+AY}(\pi^2/4I^2) \leqslant \beta_k \leqslant \frac{2AX}{AX+AY}[1 - (\pi^2/4I^2)] = m_2 \qquad (5\text{-}127)$$

and:

$$m_3 = \frac{2AY}{AX+AY}(\pi^2/4J^2) \leqslant \beta_k \leqslant \frac{2AY}{AX+AY}[1 - (\pi^2/4J^2)] = m_4 \qquad (5\text{-}128)$$

It may be that these intervals are not disjoint, in which case we may choose the interval:

$$m_5 \leqslant \beta_k \leqslant m_6 \qquad (5\text{-}129)$$

where:

$m_5 = \min(m_1, m_3)$

$m_6 = \max(m_2, m_4)$

It is the customary practice to use for β_k a geometric sequence that spans the interval (m_5, m_6), and to apply the same sequence several times until convergence is obtained. The sequence which is repeated is referred to as a cycle.

If the intervals (m_1, m_2) and (m_3, m_4) are disjoint, we may wish to choose a cycle to be the union of two geometric sequences that span the respective intervals. Not much more is known about picking parameters in this case.

Optimal parameters and convergence rate for ideal case

In the ideal case where AX and AY are constant and equal, and a square grid $(I = J)$ is used, it is possible to derive an optimal set of parameters and to estimate the associated rate of convergence.

Let s be the number of iterations (i.e., double sweeps) per cycle. Within each cycle, we use a sequence of parameters:

$$\beta_1, \beta_2, \ldots, \beta_s$$

Then the reduction in each error component over the entire cycle is given by the product:

$$\gamma_{pq}^{k+2s}/\gamma_{pq}^k = \prod_{r=1}^{s} [(\beta_r - X_p)/(\beta_r + X_p)] \cdot [(\beta_r - Y_q)/(\beta_r + Y_q)] \qquad (5\text{-}130)$$

We want to choose an optimum set of parameters so as to minimize the function:

$$\left(\max_{p,q} \prod_{r=1}^{s} |(\beta_r - X_p)/(\beta_r + X_p)| \cdot |(\beta_r - Y_q)/(\beta_r + Y_q)| \right)^{1/s} \qquad (5\text{-}131)$$

This function corresponds to the *average* error reduction per double sweep. As a consequence of the idealizations that $AX = AY$ and $I = J$, X_p and Y_q span the same spectrum, so we can instead look at minimizing the simpler function:

$$\left(\max_{p} \prod_{r=1}^{s} |(\beta_r - X_p)/(\beta_r + X_p)| \right)^{2/s} \qquad (5\text{-}132)$$

Although X_p takes on discrete values, we can approach this as a continuous problem, where X is a continuous variable lying in the interval:

$$m_1 = (\pi^2/4I^2) \leqslant X \leqslant 1 - (\pi^2/4I^2) = m_2 \qquad (5\text{-}133)$$

So now we seek to minimize the function:

$$\left(\max_{m_1 \leqslant X \leqslant m_2} \prod_{r=1}^{s} |(\beta_r - X)/(\beta_r + X)| \right)^{2/s} \qquad (5\text{-}134)$$

This is a difficult problem. An exact treatement using elliptic functions, due to Wachspress and Jordon, is described by Young (1971, p. 522).

An almost optimum method of choosing the β_r is to use the geometric sequence:

$$\beta_r = m_1(m_2/m_1)^{(r-1)/(s-1)}, \quad r = 1, 2, \ldots, s \tag{5-135}$$

where $s \geq 2$. These are known as the Wachspress parameters.

To estimate the convergence rate using these parameters, we first consider a cycle of fixed length, s. Then the error reduction over the entire cycle is given by the function:

$$\psi_s = \max_{m_1 \leq X \leq m_2} P_s(X) \tag{5-136}$$

where:

$$P_s(X) = \prod_{r=1}^{s} |(\beta_r - X)/(\beta_r + X)| \tag{5-137}$$

Divide the interval (m_1, m_2) into the $s-1$ subintervals:

$$\beta_r \leq X \leq \beta_{r+1}, \quad r = 1, 2, \ldots, s-1$$

Since the factors in the product are less than unity (and since $s \geq 2$), we can write:

$$P_s(X) < |(\beta_r - X)/(\beta_r + X)| \cdot |(\beta_{r+1} - X)/(\beta_{r+1} + X)| = P'_r(X),$$
$$\beta_r \leq X \leq \beta_{r+1} \tag{5-138}$$

where $P'_r(X)$ may be considered a partial product defined only over one subinterval. We shall see that the maximum of $P'_r(X)$ is independent of r. Therefore, if we ignore the effect of the parameters other than β_r and β_{r+1}, we can write:

$$\max_{m_1 \leq X \leq m_2} P_s(X) \approx \max_{\beta_r \leq X \leq \beta_{r+1}} P'_r(X) \tag{5-139}$$

Now $P'_r(X)$ goes through zero at the ends of the interval (β_r, β_{r+1}) and it has only one extremum, at:

$$\bar{X} = \beta_r^{\frac{1}{2}} \beta_{r+1}^{\frac{1}{2}} \tag{5-140}$$

Substitution of eq. (5-140) into (5-138) gives:

$$\max_X P'_r(X) = P'_r(\bar{X}) = (\beta_{r+1}^{\frac{1}{2}} - \beta_r^{\frac{1}{2}})^2 / (\beta_{r+1}^{\frac{1}{2}} + \beta_r^{\frac{1}{2}})^2 \tag{5-141}$$

Let

$$z = (m_2/m_1)^{1/2(s-1)} \tag{5-142}$$

From eq. (5-135):

$$\beta_{r+1}/\beta_r = z^2 \tag{5-143}$$

Combining eqs. (5-136), (5-139), (5-141), and (5-143) then gives:

$$\psi_s \approx (z-1)^2/(z+1)^2 \tag{5-144}$$

as the estimate for the error reduction over the entire cycle.

The average error reduction per double sweep, corresponding to eq. (5-134), is $(\psi_s)^{2/s}$. Then the Young convergence rate, equivalent to eq. (5-52), is

$$R_{\text{ADI}} \approx -(2/s) \ln \psi_s$$

$$R_{\text{ADI}} \approx -(4/s) \ln [(z-1)/(z+1)] \tag{5-145}$$

We seek to maximize R as a function of s. From eq. (5-142):

$$s - 1 = [\ln (m_2/m_1)]/[2 \ln z] \tag{5-146}$$

As an approximation we replace $s - 1$ by s in the above formula and substitute into eq. (5-145) to give:

$$R_{\text{ADI}} \approx \frac{-8 \ln z}{\ln (m_2/m_1)} \ln [(z-1)/(z+1)] \tag{5-147}$$

Setting $dR/dz = 0$ gives:

$$(z^2 - 1) \ln [(z+1)/(z-1)] = 2z \ln z$$

which has the solution:

$$\bar{z} = (\bar{z} + 1)/(\bar{z} - 1)$$

and:

$$\bar{z} = 2^{\frac{1}{2}} + 1 \tag{5-148}$$

The optimal choice for the cycle length, obtained by substituting eqs. (5-148) and (5-133) into eq. (5-146), is then:

$$s_{\text{opt}} = 1 + \frac{\ln (m_2/m_1)}{2 \ln \bar{z}} = 1 + 1.13 \ln (2I/\pi) \tag{5-149}$$

The corresponding value for the convergence rate is:

$$R_{\text{ADI}} \approx 8(\ln \bar{z})^2/\ln(m_2/m_1)$$

Substituting eqs. (5-133) and (5-148) then gives:

$$R_{\text{ADI}} \approx \frac{6.21}{\ln (4I^2/\pi^2)} = \frac{3.11}{\ln (2I/\pi)} \tag{5-150}$$

Thus, asymptotically (i.e., for large I), the optimal rate of convergence is proportional to $(\ln I)^{-1}$, a quantity that decreases very slowly with I. This is in contrast with the SOR procedures, where we found the optimal convergence rates to be proportional to I^{-1}.

Variable AX and AY

The theory for analysis of convergence rate for A.D.I. cannot be extended easily to problems more complex than Laplace's equation or Poisson's equation in a rectangle. The attractive estimates for convergence rates obtained for the ideal case have encouraged the use of A.D.I. for more complex problems, and it has in fact been used successfully for some quite complicated problems. In many cases, however, it has been difficult to pick a satisfactory set of parameters, and the observed rates of convergence have frequently been disappointing.

In practice, where AX and AY are highly variable, eqs. (5-127)—(5-129) can serve only as rough guides to the choice of parameters. The values of m_1, m_2, m_3, and m_4 should be evaluated for all points (i, j); then m_5 and m_6 should be obtained from:

$$m_5 = \min_{i,j} (m_1, m_3)$$

$$m_6 = \max_{i,j} (m_2, m_4)$$

It turns out that m_6 always lies between 2 and a number slightly less than 1, and experience indicates that the upper end of the range for β_k has little effect on convergence. On the other hand, the lower end of the range for β_k is found to have a profound effect on convergence rate, and even on whether the iteration will converge or diverge. It is the usual practice to use eqs. (5-127)—(5-129) to make a first guess at m_5, the lower end of the range for β_k, and to use five to ten parameters in a cycle. If a trial run shows divergence, m_5 is raised sufficiently to prevent divergence. Care must be exercised not to raise m_5 too high, or the rate of convergence will slow down too much.

For a given range (m_5, m_6), it is frequently possible to prevent divergence by using more parameters in a cycle. In fact, Pearcy (1962) has shown that for any given range (m_5, m_6), it is always possible to use enough parameters (how much is enough cannot be predicted, however) to guarantee convergence.

Other alternating-direction iteration procedures

So far we have discussed only the Peaceman-Rachford iteration procedure, which can be used only for two-dimensional elliptic problems. The Douglas-Rachford procedure can also be used for two-dimensional problems. For three-dimensional elliptic problems, the Douglas-Rachford and the Brian-Douglas procedures are available; these are straightforward counterparts of the procedures described in Chapter 3 for solving parabolic problems. Even less can be done about analyzing these methods for the choice of their iteration parameters, as compared with Peaceman-Rachford iteration, even for the case of constant AX and AY. However, these procedures have been used for solving problems with variable AX and AY, using the same guidelines for picking parameters as those given above, apparently with the same degree of success.

STRONGLY IMPLICIT PROCEDURE (S.I.P.)

Approximate factorization

Another iteration method for solving elliptic problems is the "strongly implicit procedure" proposed by Stone (1968). It is one of a family of approximate factorization methods which depend upon the observation that by making relatively small changes in the coefficient matrix **A** (which comes from eq. (5-5) written at each point), one can obtain another matrix **A'** which is extremely easy to factor; in fact, the factors **L'** and **U'** contain only three elements per row. Let us rewrite eq. (5-5) in a more general form:

$$B_{ij}P_{i,j-1} + D_{ij}P_{i-1,j} + E_{ij}P_{ij} + F_{ij}P_{i+1,j} + H_{ij}P_{i,j+1} = Q_{ij} \tag{5-151}$$

Using as an example the twelve-point problem shown in Fig. 15, we can write eq. (5-151) in matrix form, as shown in Fig. 26. We postulate that the approximate factors, **L'** and **U'**, have the same sparse structure as **A**, with the additional proviso that **U'** (like **U** in the factorization used in the band algorithm presented earlier in this chapter) has ones on its diagonal. The elements forming **L'** and **U'** are otherwise unknown at this time, but we shall designate them as b_{ij}, c_{ij}, d_{ij}, e_{ij}, and f_{ij}, as shown in Figs. 27 and 28.

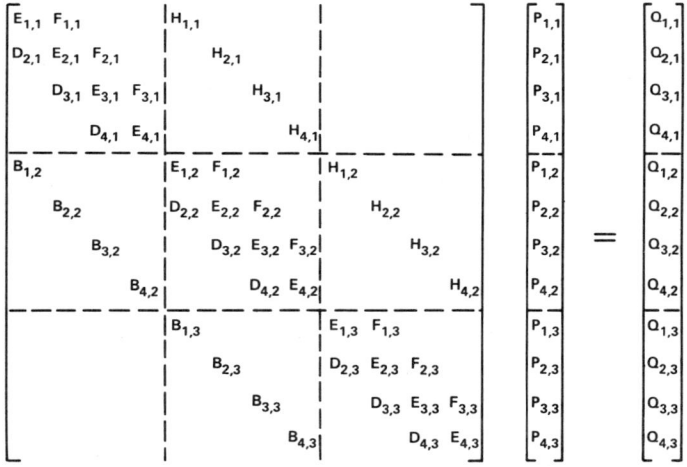

Fig. 26. Equation (5-151) in matrix form, **Ap** = **q**.

Let us carry out the matrix multiplication, **L'U'** = **A'**. We obtain the matrix shown in Fig. 29, where:

$$B'_{ij} = b_{ij} \tag{5-152a}$$

$$C'_{ij} = b_{ij}e_{i,j-1} \tag{5-152b}$$

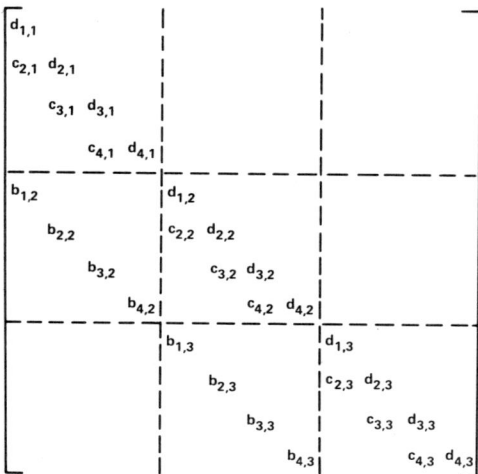

Fig. 27. Approximate lower factor, L'.

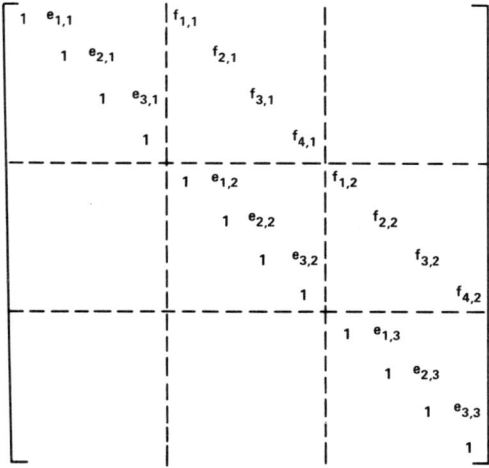

Fig. 28. Approximate upper factor, U'.

$$D'_{ij} = c_{ij} \qquad (5\text{-}152\text{c})$$
$$E'_{ij} = b_{ij}f_{i,j-1} + c_{ij}e_{i-1,j} + d_{ij} \qquad (5\text{-}152\text{d})$$
$$F'_{ij} = d_{ij}e_{ij} \qquad (5\text{-}152\text{e})$$
$$G'_{ij} = c_{ij}f_{i-1,j} \qquad (5\text{-}152\text{f})$$
$$H'_{ij} = d_{ij}f_{ij} \qquad (5\text{-}152\text{g})$$

$$\begin{bmatrix}
E'_{1,1} & F'_{1,1} & & H'_{1,1} & & & \\
D'_{2,1} & E'_{2,1} & F'_{2,1} & & G'_{2,1} & H'_{2,1} & \\
& D'_{3,1} & E'_{3,1} & F'_{3,1} & & G'_{3,1} & H'_{3,1} \\
& & D'_{4,1} & E'_{4,1} & & & G'_{4,1} & H'_{4,1} \\
\hline
B'_{1,2} & C'_{1,2} & & & E'_{1,2} & F'_{1,2} & & H'_{1,2} \\
& B'_{2,2} & C'_{2,2} & & D'_{2,2} & E'_{2,2} & F'_{2,2} & & G'_{2,2} & H'_{2,2} \\
& & B'_{3,2} & C'_{3,2} & & D'_{3,2} & E'_{3,2} & F'_{3,2} & & G'_{3,2} & H'_{3,2} \\
& & & B'_{4,2} & & & D'_{4,2} & E'_{4,2} & & & G'_{4,2} & H'_{4,2} \\
\hline
& & & & B'_{1,3} & C'_{1,3} & & & E'_{1,3} & F'_{1,3} \\
& & & & & B'_{2,3} & C'_{2,3} & & D'_{2,3} & E'_{2,3} & F'_{2,3} \\
& & & & & & B'_{3,3} & C'_{3,3} & & D'_{3,3} & E'_{3,3} & F'_{3,3} \\
& & & & & & & B'_{4,3} & & & D'_{4,3} & E'_{4,3}
\end{bmatrix}$$

Fig. 29. Product of approximate factors, $\mathbf{A}' = \mathbf{L}'\mathbf{U}'$.

We note the similarity between the structures of $\mathbf{A}'$ and $\mathbf{A}$ (Figs. 29 and 26). They do differ in one very important respect, however; $\mathbf{A}'$ contains two more diagonals, namely C'_{ij} and G'_{ij}. If we write out the equations equivalent to $\mathbf{A}'\mathbf{p} = \mathbf{q}$, we obtain:

$$B'_{ij}P_{i,j-1} + C'_{ij}P_{i+1,j-1} + D'_{ij}P_{i-1,j} + E'_{ij}P_{ij} + F'_{ij}P_{i+1,j} + G'_{ij}P_{i-1,j+1}$$
$$+ H'_{ij}P_{i,j+1} = Q_{ij} \tag{5-153}$$

and in comparing this equation with (5-151) we see the appearance of two additional terms, $C'_{ij}P_{i+1,j-1}$ and $G'_{ij}P_{i-1,j+1}$.

All of the approximate factorization methods consist of attempting to make the matrix $\mathbf{A}'$ as close as possible (in some sense) to $\mathbf{A}$, and then using $\mathbf{A}'$ as the basis for an iteration method by writing:

$$\mathbf{A}'\mathbf{p}^{k+1} = \mathbf{A}'\mathbf{p}^k - \beta(\mathbf{A}\mathbf{p}^k - \mathbf{q}) \tag{5-154}$$

where β is an iteration parameter. To reduce round-off error, we define:

$$\delta_{ij}^{k+1} = P_{ij}^{k+1} - P_{ij}^k \tag{5-155}$$

$$R_{ij}^k = Q_{ij} - B_{ij}P_{i,j-1}^k - D_{ij}P_{i-1,j}^k - E_{ij}P_{ij}^k - F_{ij}P_{i+1,j}^k - H_{ij}P_{i,j+1}^k \tag{5-156}$$

or, equivalently, in matrix form:

$$\boldsymbol{\delta}^{k+1} = \mathbf{p}^{k+1} - \mathbf{p}^k \tag{5-157}$$

$$\mathbf{r}^k = \mathbf{q} - \mathbf{A}\mathbf{p}^k \tag{5-158}$$

so that eq. (5-154) becomes:

$$A'\delta^{k+1} = \beta r^k \tag{5-159}$$

Equation (5-159) is extremely easy to solve, as we see by substituting $A' = L'U'$ and letting $v^{k+1} = U'\delta^{k+1}$. Then we have:

$$L'v^{k+1} = \beta r^k \tag{5-160}$$

$$U'\delta^{k+1} = v^{k+1} \tag{5-161}$$

From Fig. 27, we see that the forward equation (5-160) is equivalent to:

$$b_{ij}v_{i,j-1}^{k+1} + c_{ij}v_{i-1,j}^{k+1} + d_{ij}v_{ij}^{k+1} = \beta R_{ij}^k \tag{5-162}$$

while, from Fig. 28, we see that the back equation (5-161) is equivalent to:

$$\delta_{ij}^{k+1} + e_{ij}\delta_{i+1,j}^{k+1} + f_{ij}\delta_{i,j+1}^{k+1} = v_{ij}^{k+1} \tag{5-163}$$

Inasmuch as $b_{i,1}$, $c_{1,j}$, $e_{I,j}$, and $f_{i,J}$ are all zero, it follows that eq. (5-162) can be solved in order of increasing i and increasing j for successive values of v_{ij}^{k+1}, and that eq. (5-163) can then be solved in order of decreasing i and decreasing j for successive values of δ_{ij}^{k+1}. The iteration is then completed by evaluating:

$$P_{ij}^{k+1} = P_{ij}^k + \delta_{ij}^{k+1} \tag{5-164}$$

The big question remains, then, of how to specify the elements of L' and U', so as to make this iteration as rapid as possible. There are (for interior points) five unknown elements per grid point to be determined, namely b_{ij}, c_{ij}, d_{ij}, e_{ij}, and f_{ij}. However, the seven relations in eq. (5-152) must all be satisfied, in order for A' to be factorable into factors L' and U' of the form shown in Figs. 27 and 28. Thus, of the seven coefficients B'_{ij}, C'_{ij}, D'_{ij}, E'_{ij}, F'_{ij}, G'_{ij}, and H'_{ij}, only five of them can be arbitrary, leaving two of the seven coefficients to be determined from the other five by use of the seven relations in eq. (5-152).

We note in passing that if all seven of the coefficients B'_{ij} through H'_{ij} were arbitrary, we would merely set $C'_{ij} = G'_{ij} = 0$, and $B'_{ij} = B_{ij}$, $D'_{ij} = D_{ij}$, etc., and we would be finished. That we cannot do this, of course, reflects the fact that A cannot be factored into such simple factors.

This is the point at which various approximate factorization methods differ. We now consider several methods of choosing the coefficients, in order of increasing complexity.

Choice of Coefficients

Simple method
The simplest scheme is merely to set:

$$B'_{ij} = B_{ij}, \quad D'_{ij} = D_{ij}, \quad E'_{ij} = E_{ij}, \quad F'_{ij} = F_{ij}, \quad H'_{ij} = H_{ij}$$

leaving C'_{ij} and G'_{ij} to be determined from the five other coefficients. The iteration scheme so defined converges very slowly, however, compared with other methods.

Method of Dupont, Kendall and Rachford

Dupont, Kendall and Rachford (1968) propose the following. They take:

$$B'_{ij} = B_{ij}, \quad D'_{ij} = D_{ij}, \quad F'_{ij} = F_{ij}, \quad H'_{ij} = H_{ij} \tag{5-165}$$

and modify the term on the main diagonal:

$$E'_{ij} = E_{ij} - C'_{ij} - G'_{ij} + \alpha \tag{5-166}$$

where α is an iteration parameter. From eq. (5-152) we see that:

$$C'_{ij} = B'_{ij} e_{i,j-1} = B_{ij} e_{i,j-1}$$

$$G'_{ij} = D'_{ij} f_{i-1,j} = D_{ij} f_{i-1,j}$$

and since the factorization algorithm is carried out in order of increasing i and increasing j, we see that it is a simple matter to calculate, in order, b_{ij}, c_{ij}, d_{ij}, e_{ij} and f_{ij}. If we substitute eqs. (5-165) and (5-166) into eq. (5-153) to get:

$$B_{ij} P_{i,j-1} + C'_{ij}(P_{i+1,j-1} - P_{ij}) + D_{ij} P_{i-1,j} + (E_{ij} + \alpha) P_{ij} + F_{ij} P_{i+1,j}$$
$$+ G'_{ij}(P_{i-1,j+1} - P_{ij}) + H_{ij} P_{i,j+1} = Q_{ij} \tag{5-167}$$

and compare this equation with eq. (5-151), we see that Dupont, Kendall and Rachford are attempting to reduce the effect of the $C'_{ij} P_{i+1,j-1}$ term by subtracting $C'_{ij} P_{ij}$ and the effect of the $G'_{ij} P_{i-1,j+1}$ term by subtracting $G'_{ij} P_{ij}$. The iteration parameter α is added to accelerate convergence.

Method of Stone (S.I.P.)

The "strongly implicit procedure" of Stone (1968) appears to be the most effective of all. Stone noted that the values of $P_{i+1,j-1}$ and $P_{i-1,j+1}$ could be approximated (for smoothly varying P) by:

$$P_{i+1,j-1} \approx P_{i+1,j} + P_{i,j-1} - P_{ij}$$

$$P_{i-1,j+1} \approx P_{i-1,j} + P_{i,j+1} - P_{ij}$$

Hence he writes, instead of eq. (5-167):

$$B_{ij} P_{i,j-1} + C'_{ij}[P_{i+1,j-1} - \alpha_k(P_{i+1,j} + P_{i,j-1} - P_{ij})] + D_{ij} P_{i-1,j} + E_{ij} P_{ij}$$
$$+ F_{ij} P_{i+1,j} + G'_{ij}[P_{i-1,j+1} - \alpha_k(P_{i-1,j} + P_{i,j+1} - P_{ij})] + H_{ij} P_{i,j+1} = Q_{ij} \tag{5-168}$$

where α_k is an iteration parameter. The subscript k on α indicates the possibility of changing that parameter each iteration. Comparing with eq. (5-153), we see that:

$$B'_{ij} = B_{ij} - \alpha_k C'_{ij} \tag{5-169a}$$
$$D'_{ij} = D_{ij} - \alpha_k G'_{ij} \tag{5-169b}$$
$$E'_{ij} = E_{ij} + \alpha_k C'_{ij} + \alpha_k G'_{ij} \tag{5-169c}$$
$$F'_{ij} = F_{ij} - \alpha_k C'_{ij} \tag{5-169d}$$
$$H'_{ij} = H_{ij} - \alpha_k G'_{ij} \tag{5-169e}$$

Now, combine eq. (5-169a) with eqs. (5-152a) and (5-125b) to obtain:
$$b_{ij} = B_{ij} - \alpha_k(b_{ij}e_{i,j-1})$$
or:
$$b_{ij} = B_{ij}/(1 + \alpha_k e_{i,j-1})$$

Combine eq. (5-169b) with eqs. (5-152c) and (5-152f) to obtain:
$$c_{ij} = D_{ij} - \alpha_k(c_{ij}f_{i-1,j})$$
or:
$$c_{ij} = D_{ij}/(1 + \alpha_k f_{i-1,j})$$

The S.I.P. factorization algorithm can now be written in complete detail:
$$b_{ij} = B_{ij}/(1 + \alpha_k e_{i,j-1}) \tag{5-170a}$$
$$c_{ij} = D_{ij}/(1 + \alpha_k f_{i-1,j}) \tag{5-170b}$$
$$d_{ij} = E_{ij} + \alpha_k b_{ij}e_{i,j-1} + \alpha_k c_{ij}f_{i-1,j} - b_{ij}f_{i,j-1} - c_{ij}e_{i-1,j} \tag{5-170c}$$
$$e_{ij} = (F_{ij} - \alpha_k b_{ij}e_{i,j-1})/d_{ij} \tag{5-170d}$$
$$f_{ij} = (H_{ij} - \alpha_k c_{ij}f_{i-1,j})/d_{ij} \tag{5-170e}$$

Now, inasmuch as $B_{i,1}$ and $D_{1,j}$ are all zero, it follows that these five equations can be solved in order at each point, if these points are taken in order of increasing i and increasing j.

Dupont, Kendall, and Rachford (1968) were able to analyze theoretically the convergence of their method. They found that varying β in eqs. (5-154)–(5-162) can help to accelerate convergence. On the other hand, it has not been possible to develop any convergence analysis for the strongly implicit procedure. Rather, it was necessary to develop the strategy for picking S.I.P. parameters and to demonstrate the superiority of S.I.P. empirically. Stone found no advantage in choosing β different from one, thus leaving only one iteration parameter, α_k, to be chosen in S.I.P.

Varying α_k through a cycle of values is extremely important. It has been found that $(1 - \alpha_k)$ should be computed in the same way that β_k is computed for A.D.I. The minimum value of α_k is not critical and may be taken as 0. The maximum value of α_k is quite critical (though the method is not nearly as sensitive as A.D.I. to this choice) and it has been found satisfactory to calculate it from:

$$1 - \alpha_{max} = \min_{ij}\left[\frac{\pi^2}{2I^2\left(1+\frac{AY}{AX}\right)}, \frac{\pi^2}{2J^2\left(1+\frac{AX}{AY}\right)}\right] \quad (5\text{-}171)$$

where AX are the coefficients D_{ij} and F_{ij}, and AY are the coefficients B_{ij} and H_{ij}. The different values of $(1-\alpha_k)$ in the cycle are chosen to be spaced geometrically between 1 and $(1-\alpha_{max})$; that is:

$$1 - \alpha_r = (1-\alpha_{max})^{(r-1)/(s-1)}, \quad r = 1, 2, \ldots, s \quad (5\text{-}172)$$

where s is the number of parameters in the cycle. If the problem has compressibility (so that $|E_{ij}| > |B_{ij} + D_{ij} + F_{ij} + H_{ij}|$), then an $\alpha_k = 1$ should be included in the cycle.

Stone found that changing the order of points in alternate iterations greatly improves the speed of convergence. That is, in every second iteration, the factorization and forward solution should proceed in the order of decreasing j and increasing i, while the back solution proceeds in the order of decreasing i and increasing j. This has the effect of bringing in two different points, $(i-1, j-1)$ and $(i+1, j+1)$ into eq. (5-153), rather than the points $(i+1, j-1)$ and $(i-1, j+1)$.

The S.I.P. algorithm presented above is, of course, limited to the solution of two-dimensional problems. Weinstein et al. (1969) have extended the algorithm to permit the application of the strongly implicit procedure to the solution of three-dimensional elliptic problems.

SUMMARY

Experience shows that for the very simple, or ideal problems, where AX and AY are almost constant, A.D.I. is the fastest iteration method. For widely varying AX and AY, for complex-shaped reservoir regions, and/or with very high or low ratios of AY to AX, the convergence of A.D.I. deteriorates badly and, in some cases, it has not been possible to choose iteration parameters to make it converge. In such difficult cases, S.I.P. converges only a little slower than it does for ideal problems. Further, the use of eqs. (5-171) and (5-172) to choose iteration parameters is relatively reliable, making the search for good iteration parameters for S.I.P. much easier than for other iteration methods.

Line relaxation methods with additive corrections also constitute a viable choice for the solution of difficult problems. Because of the great variety of situations that can be encountered in reservoir simulation, at the present time the most reliable simulators are those which offer the user the choice among direct solution, S.I.P., and LSORC, for the solution of the elliptic problems encountered in the simulation.

NOMENCLATURE

Units

Some of the symbols listed below are defined in terms of SI units, kg (kilogram), m (meter), s (second), N (newton = kg · m/s^2) and Pa (pascal = N/m^2).

Symbols

a	$= (\beta + \omega - 1)/\beta$
$\mathbf{A}$	= coefficient matrix
$\mathbf{A}'$	= approximate coefficient matrix
AE_{ij}	= "east" coefficient, corresponding to $AX_{i+\frac{1}{2},j}$
AN_{ij}	= "north" coefficient, corresponding to $AY_{i,j+\frac{1}{2}}$
AS_{ij}	= "south" coefficient, corresponding to $AY_{i,j-\frac{1}{2}}$
$AX_{i+\frac{1}{2},j}$	$= KX_{i+\frac{1}{2},j}\Delta y_j/(x_{i+1} - x_i) =$ transmissibility in x-direction [m^2/Pa · s]
AW_{ij}	= "west" coefficient, corresponding to $AX_{i-\frac{1}{2},j}$
$AY_{i,j+\frac{1}{2}}$	$= KY_{i,j+\frac{1}{2}}\Delta x_i/(y_{j+1} - y_j) =$ transmissibility in y-direction [m^2/Pa · s]
b_{ij}	= element of lower approximate factor $\mathbf{L}'$
b_{rs}	= element of band matrix $\mathbf{B}$
B_{ij}	= same as $-AS_{ij}$ = element of coefficient matrix $\mathbf{A}$
B'_{ij}	= element of approximate coefficient matrix $\mathbf{A}'$
$\mathbf{B}$	= band matrix
c_{ij}	= element of lower approximate factor $\mathbf{L}'$
C'_{ij}	= element of approximate coefficient matrix $\mathbf{A}'$
d_{ij}	= diagonal element of coefficient matrix $\mathbf{A}$, or of lower approximate factor $\mathbf{L}'$
D_{ij}	= same as $-AW_{ij}$ = element of coefficient matrix $\mathbf{A}$
D'_{ij}	= element of approximate coefficient matrix $\mathbf{A}'$
$\mathbf{D}$	= diagonal matrix
e_{ij}	= element of upper approximate factor $\mathbf{U}'$
$\mathbf{e}^k$	= error vector, whose components are E_{ij}^k
E_{ij}	= same as d_{ij} = diagonal element of coefficient matrix $\mathbf{A}$
E_{ij}^k	= error in solution for P_{ij} after k iterations
E'_{ij}	= diagonal element of approximate coefficient matrix $\mathbf{A}'$
f_{ij}	= element of upper approximate factor $\mathbf{U}'$
F_{ij}	= same as $-AE_{ij}$ = element of coefficient matrix $\mathbf{A}$
F'_{ij}	= element of approximate coefficient matrix $\mathbf{A}'$
g_r	= intermediate variable in direct solution algorithm
$\mathbf{g}$	= vector whose elements are g_r
G'_{ij}	= element of approximate coefficient matrix $\mathbf{A}'$
$\mathbf{G}$	= iteration matrix
H_{ij}	= same as $-AN_{ij}$ = element of coefficient matrix $\mathbf{A}$
H'_{ij}	= element of approximate coefficient matrix $\mathbf{A}'$
$\mathbf{H}$	= "horizontal" portion of coefficient matrix $\mathbf{A}$; used in A.D.I.
i	= index of grid points or blocks in x-direction
i'	= half-index in x-direction, equals $i - \frac{1}{2}$
i	$= (-1)^{\frac{1}{2}}$
I	= number of grid points or blocks in x-direction
$\mathbf{I}$	= identity matrix
j	= index of grid points or blocks in y-direction
j'	= half-index in y-direction, equals $j - \frac{1}{2}$
J	= number of grid points or blocks in y-direction

KX	=	mobility in x-direction [m^2/Pa · s]
KY	=	mobility in y-direction [m^2/Pa · s]
l_{rs}	=	element of lower triangular matrix **L**
L	=	length of reservoir (in x-direction) [m]
L	=	lower triangular factor, used in direct solution
L$'$	=	lower approximate factor of coefficient matrix **A**
m	=	limits on range of A.D.I. parameters
M	=	general matrix
N	=	number of equations, equals $I \cdot J$
p	=	pressure [Pa]
p	=	index in Fourier expansion for error
p	=	solution vector, whose components are P_{ij}
P_{ij}	=	difference solution for pressure p [Pa]
$P_s(X)$	=	product of functions of X, defined by eq. (5-137)
$P'_r(X)$	=	partial product of functions of X, defined by eq. (5-138)
q	=	volume rate of injection (or production, if negative) per unit volume of reservoir [s^{-1}]
q	=	index in Fourier expansion for error
q	=	input-output vector, whose components are Q_{ij}
Q_{ij}	=	volume rate of injection (or production, if negative) for grid block (i, j) per unit thickness, equals $q_{ij}\Delta x_i \Delta y_j$ [m^2/s]
r	=	residual vector, whose components are R_{ij}
R	=	$R(\mathbf{G})$ = Young's rate of convergence of iteration, defined by eq. (5-52)
R_{ij}	=	residual, defined by eq. (5-25) [m^2/s]
s	=	number of iterations (or double sweeps) per cycle
S	=	normalizing factor used in A.D.I. [m^2/Pa · s]
u_{rs}	=	element of upper triangular factor **U**
U	=	upper triangular factor, used in direct solution
U$'$	=	upper approximate factor of coefficient matrix **A**
v_{ij}	=	intermediate variable in approximate factorization
v	=	vector whose components are v_{ij}
v$_{pq}$	=	eigenvector
V_{ij}^{pq}	=	element of eigenvector **v**$_{pq}$
V	=	"vertical" portion of coefficient matrix **A**; used in A.D.I.
W	=	width of reservoir (in y-direction) [m]
W	=	width of band matrix
W'	=	half-width of band matrix, equals $(W-1)/2$
x	=	distance [m]
X	=	continuous variable over interval (m_1, m_2)
X_p	=	$2AX(AX+AY)^{-1} \sin^2(p\pi/2I)$
y	=	distance [m]
Y_q	=	$2AY(AX+AY)^{-1} \sin^2(q\pi/2J)$
z	=	$(m_2/m_1)^{1/2(s-1)}$
α	=	iteration parameter
α_i	=	column correction, used in method of additive corrections [Pa]
β	=	overrelaxation parameter in SOR methods; also iteration parameter in A.D.I. and in iterations using approximate factorization
γ_{pq}^k	=	coefficient in Fourier expansion for error
δ_{ij}^{k+1}	=	change in P_{ij} in kth iteration, equals $P_{ij}^{k+1} - P_{ij}^k$ [Pa]
δ_x^2	=	second-difference operator in x-direction, defined by eq. (5-45)
δ_y^2	=	second-difference operator in y-direction, defined by eq. (5-46)
δ^k	=	solution change vector, whose elements are δ_{ij}^k

Δx_i = $x_{i+\frac{1}{2}} - x_{i-\frac{1}{2}}$ [m]
Δy_j = $y_{j+\frac{1}{2}} - y_{j-\frac{1}{2}}$ [m]
λ = eigenvalue of matrix $D^{-1}A$
Γ = special diagonal matrix
μ = eigenvalue of Jacobi iteration matrix
ω = eigenvalue of SOR iteration matrix

Subscripts and superscripts

ADI	refers to alternating-direction iteration
c	refers to corrected iterate in method of additive corrections
C	refers to iterative method with additive correction
i	index in x-direction
j	index in y-direction
J	refers to Jacobi iteration method
k	iteration count
k	summation index in direct solution algorithm
L	refers to line version of an iteration method
opt	optimum
p	index in Fourier expansion for error
q	index in Fourier expansion for error
r	row index of a matrix
s	column index of a matrix
SOR	refers to successive overrelaxation
∞	refers to solution after convergence
*	refers to intermediate solution (before extrapolation) in SOR methods

CHAPTER 6

NUMERICAL SOLUTION OF TWO-PHASE FLOW PROBLEMS

INTRODUCTION

In this final chapter, we shall examine several numerical methods for the solution of multidimensional, multiphase flow problems. For the sake of clarity and brevity of presentation, the discussion will be limited to the case of horizontal, two-dimensional flow of two incompressible phases with no mass transfer between phases. Of course, current practical resevoir simulators are much more general, in that they include the effects of compressibility, gravity, and some mass transfer (using at least a black-oil system). Further, modern simulators frequently deal with three phases and are increasingly used to solve problems in three dimensions.

The numerical methods used in these more general simulators involve, for the most part, extensions of methods which can be adequately, and certainly more clearly, described in terms of the simpler two-phase incompressible case. It turns out that compressible problems are easier to solve numerically than incompressible problems. Paradoxically, the equations that describe incompressible flow problems are simpler than those which describe compressible ones, even though incompressible problems present more numerical difficulties and are more interesting from a mathematical point of view. Furthermore, inasmuch as any reliable simulator should be able to deal with any level of compressibility, down to zero, the numerical procedures must be based on methods which work for the incompressible case.

DIFFERENTIAL EQUATIONS

Basic equations in terms of phase pressures

The assumptions of horizontal flow of incompressible fluids imply that D, ρ_n, and ρ_w are constant. In addition, to simplify matters further, we ignore the possible variation of porosity, ϕ, with pressure, and of the geometric factor, α, with position. Then eqs. (1-40) become:

$$\nabla \cdot [(Kk_{rn}/\mu_n)\nabla p_n] + Q_n = \phi \frac{\partial S_n}{\partial t} \tag{6-1}$$

$$\nabla \cdot [(Kk_{rw}/\mu_w)\nabla p_w] + Q_w = \phi \frac{\partial S_w}{\partial t} \tag{6-2}$$

where $Q_n = q_n/\rho_n$ and $Q_w = q_w/\rho_w$ are the *volumetric* rates of injection per unit of reservoir volume.

Recall the definitions for phase mobilities (eqs. 1-48):

$$\lambda_n = Kk_{rn}/\mu_n \tag{6-3}$$

$$\lambda_w = Kk_{rw}/\mu_w \tag{6-4}$$

Then eqs. (6-1)—(6-2) may be written:

$$\nabla \cdot (\lambda_n \nabla p_n) + Q_n = \phi \frac{\partial S_n}{\partial t} \tag{6-5}$$

$$\nabla \cdot (\lambda_w \nabla p_w) + Q_w = \phi \frac{\partial S_w}{\partial t} \tag{6-6}$$

If we expand the accumulation terms, we get:

$$\frac{\partial S_n}{\partial t} = -\frac{\partial S_w}{\partial t} = \frac{dS_n}{dp_c}\frac{\partial p_c}{\partial t} = S'\left(\frac{\partial p_n}{\partial t} - \frac{\partial p_w}{\partial t}\right)$$

where $S' = dS_n/dp_c$ is considered to be a unique function of S_n (or of p_c). Then eqs. (6-5)—(6-6) may be written:

$$\nabla \cdot (\lambda_n \nabla p_n) + Q_n = \phi S'\left(\frac{\partial p_n}{\partial t} - \frac{\partial p_w}{\partial t}\right) \tag{6-7}$$

$$\nabla \cdot (\lambda_w \nabla p_w) + Q_w = \phi S'\left(\frac{\partial p_w}{\partial t} - \frac{\partial p_n}{\partial t}\right) \tag{6-8}$$

Note that S', as defined above, is always positive.

Alternative equations

An alternative form of eqs. (6-7)—(6-8) is of interest. Define the total mobility and the difference of mobilities by:

$$\lambda_t = \lambda_n + \lambda_w \tag{6-9}$$

$$\lambda_d = \lambda_n - \lambda_w \tag{6-10}$$

In addition, average pressure and capillary pressure are defined by:

$$p_{avg} = (p_n + p_w)/2$$

$$p_c = p_n - p_w$$

By adding and subtracting eqs. (6-7) and (6-8), we obtain:

$$\nabla \cdot (\lambda_t \nabla p_{avg}) + \tfrac{1}{2}\nabla \cdot (\lambda_d \nabla p_c) + Q_n + Q_w = 0 \tag{6-11}$$

$$\tfrac{1}{2}\nabla \cdot (\lambda_t \nabla p_c) + \nabla \cdot (\lambda_d \nabla p_{avg}) + Q_n - Q_w = 2\phi S' \frac{\partial p_c}{\partial t} \tag{6-12}$$

Equation (6-11) could have been obtained directly from the pressure equation (1-49) by using the simplifying assumptions of constant D, ρ_n, ρ_w, α, and ϕ. While eq. (6-12) can be considered to be a saturation equation, its relationship to the saturation equation (1-58) is much less direct.

By making use of the concept of total velocity, however, we can use the Buckley-Leverett form of the saturation equation (1-58). The individual phase velocities are, from eqs. (1-38) and the assumption of horizontal flow:

$$\vec{v}_n = -\lambda_n \nabla p_n \tag{6-13}$$

$$\vec{v}_w = -\lambda_w \nabla p_w \tag{6-14}$$

and the total velocity is:

$$\vec{v}_t = \vec{v}_n + \vec{v}_w \tag{6-15}$$

With the simplifying assumptions of constant D, ρ_n, ρ_w, α, and ϕ, together with the substitution of eqs. (1-55) and (1-56), the Buckley-Leverett form of the saturation equation (1-58) becomes:

$$-\nabla \cdot (f_w \lambda_n \nabla p_c) - \nabla \cdot (f_w \vec{v}_t) + Q_w = \phi \frac{\partial S_w}{\partial t} \tag{6-16}$$

DIFFERENCE NOTATION

Difference operators

To simplify the writing of the difference equations, let us reintroduce some difference notation. For each interval of the grid, we define phase mobility coeffients, such as:

$$LNX_{i+\frac{1}{2},j,m} \approx (\lambda_n)_{i+\frac{1}{2},j,m}$$

$$LWY_{i,j+\frac{1}{2},m} \approx (\lambda_w)_{i,j+\frac{1}{2},m}$$

Then difference operators which serve as the analogs of the derivatives:

$$\frac{\partial}{\partial x}\left(\lambda_n \frac{\partial p_n}{\partial x}\right) \quad \text{and} \quad \frac{\partial}{\partial y}\left(\lambda_n \frac{\partial p_n}{\partial y}\right)$$

are:

$\Delta_x(LN_m \Delta_x PN_n)$

$$= \frac{LNX_{i+\frac{1}{2},j,m} \dfrac{PN_{i+1,j,n} - PN_{ijn}}{x_{i+1} - x_i} - LNX_{i-\frac{1}{2},j,m} \dfrac{PN_{ijn} - PN_{i-1,j,n}}{x_i - x_{i-1}}}{\Delta x_i} \qquad (6\text{-}17)$$

and:

$\Delta_y(LN_m \Delta_y PN_n)$

$$= \frac{LNY_{i,j+\frac{1}{2},m} \dfrac{PN_{i,j+1,n} - PN_{ijn}}{y_{j+1} - y_j} - LNY_{i,j-\frac{1}{2},m} \dfrac{PN_{ijn} - PN_{i,j-1,n}}{y_j - y_{j-1}}}{\Delta y_j} \qquad (6\text{-}18)$$

where:

$$\Delta x_i = x_{i+\frac{1}{2}} - x_{i-\frac{1}{2}} \qquad (6\text{-}19)$$

$$\Delta y_j = y_{j+\frac{1}{2}} - y_{j-\frac{1}{2}} \qquad (6\text{-}20)$$

Note the similarity of eqs. (6-17) and (6-18) with (2-19) and (2-20). Finally, it is convenient to define a combined, or two-dimensional, operator:

$$\Delta(LN_m \Delta PN_n) = \Delta_x(LN_m \Delta_x PN_n) + \Delta_y(LN_m \Delta_y PN_n) \qquad (6\text{-}21)$$

which is, of course, the difference analog of the differential operator $\nabla \cdot (\lambda_n \nabla p_n)$.

Injection-production terms

We need to consider various forms of the injection (or production) terms. Recall that Q_n and Q_w are volumetric rates of injection (if positive) or of production (if negative) per unit volume of reservoir. Let:

$$\overline{QN}_{ij} = \int_{x_{i-\frac{1}{2}}}^{x_{i+\frac{1}{2}}} \int_{y_{j-\frac{1}{2}}}^{y_{j+\frac{1}{2}}} Q_n \alpha \, dy \, dx \quad \text{and:} \quad \overline{QW}_{ij} = \int_{x_{i-\frac{1}{2}}}^{x_{i+\frac{1}{2}}} \int_{y_{j-\frac{1}{2}}}^{y_{j+\frac{1}{2}}} Q_w \alpha \, dy \, dx$$

be the total volumetric rates associated with the grid point (i, j). Actually, injection or production rates are usually not distributed continuously over the area occupied by the block surrounding each grid point but, rather, are associated with wells which may be considered to be mathematical point sources and sinks. Thus, $\overline{QN}_{ij}$ and $\overline{QW}_{ij}$ represent volumetric rates (i.e., volume per unit time) associated with a well located at the point (i, j).

In replacing differential equations with their difference analogs, then, we will replace Q_n and Q_w by:

$$QN_{ij} = \overline{QN}_{ij}/(\alpha \Delta x_i \Delta y_j) \qquad (6\text{-}22)$$

$$QW_{ij} = \overline{QW}_{ij}/(\alpha \Delta x_i \Delta y_j) \qquad (6\text{-}23)$$

Interval absolute permeabilities

We allow for the possibility that the absolute permeability is anisotropic; that is, that K_x is not necessarily equal to K_y. For each grid interval, we define:

$$KX_{i+\frac{1}{2},j} \approx (K_x)_{i+\frac{1}{2},j} \quad \text{and} \quad KY_{i,j+\frac{1}{2}} \approx (K_y)_{i,j+\frac{1}{2}}$$

These quantities may be input directly by the user, or they may be evaluated from grid-point values of K_x and K_y which are input by the user. In the latter case, the arithmetic average is usually used, e.g.:

$$KX_{i+\frac{1}{2},j} = [(K_x)_{ij} + (K_x)_{i+1,j}]/2$$

although sometimes a geometric average is used, namely:

$$KX_{i+\frac{1}{2},j} = (K_x)_{ij}^{\frac{1}{2}}(K_x)_{i+1,j}^{\frac{1}{2}}$$

Mobility weighting

For each grid interval, the mobility coefficients can now be evaluated in more detail. We write:

$$LNX_{i+\frac{1}{2},j,m} = KX_{i+\frac{1}{2},j}[(W)(k_{rn}/\mu_n)_{ijm} + (1-W)(k_{rn}/\mu_n)_{i+1,j,m}] \qquad (6\text{-}24)$$

The quantity W is the weighting factor on the mobility. It is analogous to the distance-weighting factor discussed in Chapter 4 in connection with the solution of the nonlinear convection equation (4-1). $W = \frac{1}{2}$ corresponds to *midpoint* weighting. The more commonly used alternative is *upstream* weighting, wherein W is chosen to be 0 or 1, depending on the direction of the potential gradient in that phase for that interval. For this example, if flow is from left to right, we use the mobility at the left (upstream) end of the interval and pick $W = 1$.

Similar definitions are made for:

$$LNY_{i,j+\frac{1}{2},m}, \quad LWX_{i+\frac{1}{2},j,m}, \quad \text{and} \quad LWY_{i,j+\frac{1}{2},m}$$

Note that, at this point, we have not specified m and therefore have not committed ourselves as to the time level at which the mobilities (k_r/μ) are to be evaluated.

Some special combinations

In analogy with eqs. (6-9) and (6-10), we can define the following terms:

$$LTX_{i+\frac{1}{2},j,m} = LNX_{i+\frac{1}{2},j,m} + LWX_{i+\frac{1}{2},j,m} \qquad (6\text{-}25)$$

$$LDX_{i+\frac{1}{2},j,m} = LNX_{i+\frac{1}{2},j,m} - LWX_{i+\frac{1}{2},j,m} \qquad (6\text{-}26)$$

Similar definitions hold for $LTY_{i,j+\frac{1}{2},m}$ and $LDY_{i,j+\frac{1}{2},m}$. In addition, we use PA_{ij} and PC_{ij} to refer to the difference solutions for p_{avg} and p_c, respectively. Then we can proceed to define operators, such as:

$$\Delta_x(LT_m \Delta_x PA_n), \quad \Delta_y(LD_m \Delta_y PC_n)$$

etc., as well as the combined two-dimensional operators, such as:

$$\Delta(LT_m \Delta PA_n), \quad \Delta(LD_m \Delta PC_n)$$

etc. These will be useful in setting up difference analogs of eqs. (6-11) and (6-12).

SIMULTANEOUS NUMERICAL SOLUTION

Explicit difference equations

First, let us consider an explicit difference analog of eqs. (6-11)–(6-12):

$$\Delta(LT_n \Delta PA_n) + \tfrac{1}{2}\Delta(LD_n \Delta PC_n) + (QN + QW)_{ijn} = 0$$

$$\tfrac{1}{2}\Delta(LT_n \Delta PC_n) + \Delta(LD_n \Delta PA_n) + (QN - QW)_{ijn}$$
$$= 2(\phi S'/\Delta t)_{ijn}(PC_{i,j,n+1} - PC_{ijn})$$

We can solve directly for $PC_{i,j,n+1}$. But there is no explicit way to solve for $PA_{i,j,n+1}$!

If we write an explicit difference analog of eqs. (6-7)–(6-8), we have:

$$\Delta(LN_n \Delta PN_n) + QN_{ijn} = (\phi S'/\Delta t)_{ijn}(PN_{i,j,n+1} - PN_{ijn} - PW_{i,j,n+1} + PW_{ijn})$$
(6-27)

$$\Delta(LW_n \Delta PW_n) + QW_{ijn} = (\phi S'/\Delta t)_{ijn}(PW_{i,j,n+1} - PW_{ijn} - PN_{i,j,n+1} + PN_{ijn})$$
(6-28)

It is clear that these equations are inconsistent. The right-hand side of eq. (6-27) is the negative of the right-hand side of (6-28); two separate answers can thus be obtained for the explicit calculation of the change in capillary pressure, i.e., for:

$$PC_{i,j,n+1} - PC_{ijn} = (PN - PW)_{i,j,n+1} - (PN - PW)_{ijn}$$

while it is not possible to calculate the changes in PW or PN themselves. Hence, we must rule out any such simple approach as these totally explicit procedures.

Alternating-direction implicit procedure

Next, we consider an alternating-direction *implicit* procedure that is an extension of the A.D.I. method of eqs. (3-54), which was used to solve a parabolic problem in one dependent variable. For this purpose, eqs.

(6-7)—(6-8) are exactly equivalent to eqs. (6-11)—(6-12). It will be more convenient for analysis to work with the latter pair. The A.D.I. difference equations are, then, for the horizontal sweep:

$$\Delta_x(LT_n\Delta_x PA_{n+1}) + \Delta_y(LT_n\Delta_y PA_n) + \tfrac{1}{2}\Delta_x(LD_n\Delta_x PC_{n+1}) + \tfrac{1}{2}\Delta_y(LD_n\Delta_y PC_n)$$
$$+ (QN + QW)_{ijn} = 0 \tag{6-29}$$

$$\tfrac{1}{2}\Delta_x(LT_n\Delta_x PC_{n+1}) + \tfrac{1}{2}\Delta_y(LT_n\Delta_y PC_n) + \Delta_x(LD_n\Delta_x PA_{n+1}) + \Delta_y(LD_n\Delta_y PA_n)$$
$$+ (QN - QW)_{ijn} = 2(\phi S'/\Delta t)_{ij}(PC_{i,j,n+1} - PC_{ijn}) \tag{6-30}$$

and for the vertical sweep:

$$\Delta_x(LT_n\Delta_x PA_{n+1}) + \Delta_y(LT_n\Delta_y PA_{n+2}) + \tfrac{1}{2}\Delta_x(LD_n\Delta_x PC_{n+1})$$
$$+ \tfrac{1}{2}\Delta_y(LD_n\Delta_y PC_{n+2}) + (QN + QW)_{ijn} = 0 \tag{6-31}$$

$$\tfrac{1}{2}\Delta_x(LT_n\Delta_x PC_{n+1}) + \tfrac{1}{2}\Delta_y(LT_n\Delta_y PC_{n+2}) + \Delta_x(LD_n\Delta_x PA_{n+1})$$
$$+ \Delta_y(LD_n\Delta_y PA_{n+2}) + (QN - QW)_{ijn}$$
$$= 2(\phi S'/\Delta t)_{ij}(PC_{i,j,n+2} - PC_{i,j,n+1}) \tag{6-32}$$

For a von Neumann stability analysis, we will drop QN and QW and assume constant Δx, Δy, ϕ, S', $LTX = LTY$, and $LDX = LDY$. Multiply each equation through by $(\Delta t/2\phi S')$. Let:

$$a = LTX(\Delta t/2\phi S') = LTY(\Delta t/2\phi S') \tag{6-33}$$
$$b = LDX(\Delta t/2\phi S') = LDY(\Delta t/2\phi S') \tag{6-34}$$

Then the horizontal sweep simplifies to:

$$\Delta_x^2(aPA_{n+1} + \tfrac{1}{2}bPC_{n+1}) + \Delta_y^2(aPA_n + \tfrac{1}{2}bPC_n) = 0 \tag{6-35}$$
$$\Delta_x^2(bPA_{n+1} + \tfrac{1}{2}aPC_{n+1}) + \Delta_y^2(bPA_n + \tfrac{1}{2}aPC_n) = PC_{i,j,n+1} - PC_{ijn} \tag{6-36}$$

and the vertical sweep to:

$$\Delta_x^2(aPA_{n+1} + \tfrac{1}{2}bPC_{n+1}) + \Delta_y^2(aPA_{n+2} + \tfrac{1}{2}bPC_{n+2}) = 0 \tag{6-37}$$
$$\Delta_x^2(bPA_{n+1} + \tfrac{1}{2}aPC_{n+1}) + \Delta_y^2(bPA_{n+2} + \tfrac{1}{2}aPC_{n+2}) = PC_{i,j,n+2} - PC_{i,j,n+1} \tag{6-38}$$

We assume that errors of the form of eq. (3-36) satisfy these difference equations; hence we assume:

$$PA_{ijn} = \gamma^n \exp(ipi\Delta x)\exp(iqj\Delta y) \tag{6-39}$$
$$PC_{ijn} = \xi^n \exp(ipi\Delta x)\exp(iqj\Delta y) \tag{6-40}$$

Upon substituting eqs. (6-39)—(6-40) into eqs. (6-35)—(6-38) and cancelling common factors, we obtain, for the horizontal sweep:

$$-X_p(a\gamma^{n+1} + \tfrac{1}{2}b\xi^{n+1}) - Y_q(a\gamma^n + \tfrac{1}{2}b\xi^n) = 0 \tag{6-41}$$

$$-X_p(b\gamma^{n+1} + \tfrac{1}{2}a\xi^{n+1}) - Y_q(b\gamma^n + \tfrac{1}{2}a\xi^n) = \xi^{n+1} - \xi^n \tag{6-42}$$

and for the vertical sweep:

$$-X_p(a\gamma^{n+1} + \tfrac{1}{2}b\xi^{n+1}) - Y_q(a\gamma^{n+2} + \tfrac{1}{2}b\xi^{n+2}) = 0 \tag{6-43}$$

$$-X_p(b\gamma^{n+1} + \tfrac{1}{2}a\xi^{n+1}) - Y_q(b\gamma^{n+2} + \tfrac{1}{2}a\xi^{n+2}) = \xi^{n+2} - \xi^{n+1} \tag{6-44}$$

where:

$$X_p = (4/\Delta x^2) \sin^2(p\Delta x/2) \tag{6-45}$$

$$Y_q = (4/\Delta y^2) \sin^2(q\Delta y/2) \tag{6-46}$$

By multiplying eqs. (6-41) and (6-43) by $-b/a$ and adding, respectively, to eqs. (6-42) and (6-44), we obtain:

$$-\tfrac{1}{2}X_p[a-(b^2/a)]\xi^{n+1} - \tfrac{1}{2}Y_q[a-(b^2/a)]\xi^n = \xi^{n+1} - \xi^n \tag{6-47}$$

$$-\tfrac{1}{2}X_p[a-(b^2/a)]\xi^{n+1} - \tfrac{1}{2}Y_q[a-(b^2/a)]\xi^{n+2} = \xi^{n+2} - \xi^{n+1} \tag{6-48}$$

Equation (6-47) can be solved for ξ^{n+1}/ξ^n, while eq. (6-48) can be solved for ξ^{n+2}/ξ^{n+1}. Upon multiplying these two ratios together, we obtain:

$$\frac{\xi^{n+2}}{\xi^n} = \frac{1 - \tfrac{1}{2}[a-(b^2/a)]X_p}{1 + \tfrac{1}{2}[a-(b^2/a)]X_p} \cdot \frac{1 - \tfrac{1}{2}[a-(b^2/a)]Y_q}{1 + \tfrac{1}{2}[a-(b^2/a)]Y_q} \tag{6-49}$$

Since $a > 0$ and $|b| < a$, then $|\xi^{n+2}/\xi^n| < 1$, and it appears that this method is stable. However, we must also examine γ. From eqs. (6-41) and (6-43), we have:

$$a\gamma^{n+2} + \tfrac{1}{2}b\xi^{n+2} = a\gamma^n + \tfrac{1}{2}b\xi^n$$

Thus the linear combination, $a\gamma + \tfrac{1}{2}b\xi$, remains constant and does not decay, as we should expect from a stable process. Since eq. (6-49) shows ξ to be decreasing in magnitude, we should therefore expect that γ will increase somewhat in magnitude with each double time step. This means that a simultaneous alternating-direction *implicit* procedure cannot be expected to be stable.

It *is* true that if compressibility were included, then this alternating-direction implicit method will work for sufficiently large compressibility or, equivalently, for sufficiently small time step. But since we have stipulated that our simulators be sufficiently general to include small or zero compressibility, we must therefore rule out this approach also.

Simultaneous implicit procedure

Since explicit and alternating-direction *implicit* methods are not stable for incompressible two-phase flow problems, we must consider other implicit procedures. We examine now a simultaneous implicit procedure, introduced by Douglas, Peaceman and Rachford (1959), which was the first method to be used successfully for such problems in two dimensions.

Because simultaneous solution for two dependent variables is involved, eqs. (6-7)–(6-8) are exactly equivalent to eqs. (6-11)–(6-12). It is easier to analyze for stability using the latter pair, although the former pair is more convenient to use when setting up the solution procedure.

Stability analysis

The difference analog of eqs. (6-11)–(6-12) that is completely implicit in pressure is the following pair of equations:

$$\Delta(LT_n \Delta PA_{n+1}) + \tfrac{1}{2}\Delta(LD_n \Delta PC_{n+1}) + (QN + QW)_{ijn} = 0 \tag{6-50}$$

$$\tfrac{1}{2}\Delta(LT_n \Delta PC_{n+1}) + \Delta(LD_n \Delta PA_{n+1}) + (QN - QW)_{ijn}$$
$$= 2(\phi S'/\Delta t)_{ij}(PC_{i,j,n+1} - PC_{ijn}) \tag{6-51}$$

This will, of course, give rise to a formidable system of linear equations to solve, since there are two equations and two unknowns for every grid point. Let us postpone the discussion of how to solve this difference system until we examine its stability. For the stability analysis, we use the same approach as used above with the alternating-direction implicit procedure, in that we drop QN and QW and assume constant Δx, Δy, ϕ, S', $LTX = LTY$, and $LDX = LDY$. Multiply each equation through by $(\Delta t/2\phi S')$. Then eqs. (6-50)–(6-51) become:

$$\Delta_x^2(aPA_{n+1} + \tfrac{1}{2}bPC_{n+1}) + \Delta_y^2(aPA_{n+1} + \tfrac{1}{2}bPC_{n+1}) = 0 \tag{6-52}$$

$$\Delta_x^2(bPA_{n+1} + \tfrac{1}{2}aPC_{n+1}) + \Delta_y^2(bPA_{n+1} + \tfrac{1}{2}aPC_{n+1}) = PC_{i,j,n+1} - PC_{ijn} \tag{6-53}$$

Again, we assume solutions of the form (6-39) and (6-40). Upon substituting into (6-52)–(6-53) and cancelling common factors, we obtain:

$$-(X_p + Y_q)(a\gamma^{n+1} + \tfrac{1}{2}b\xi^{n+1}) = 0 \tag{6-54}$$

$$-(X_p + Y_q)(b\gamma^{n+1} + \tfrac{1}{2}a\xi^{n+1}) = \xi^{n+1} - \xi^n \tag{6-55}$$

Multiplying eq. (6-54) by $-b/a$ and adding to (6-55) gives:

$$-\tfrac{1}{2}(X_p + Y_q)[a - (b^2/a)]\xi^{n+1} = \xi^{n+1} - \xi^n$$

or:

$$\frac{\xi^{n+1}}{\xi^n} = \frac{1}{1 + \tfrac{1}{2}[a - (b^2/a)](X_p + Y_q)} < 1 \tag{6-56}$$

From eq. (6-54) we have:

$$\gamma^{n+1} = -\tfrac{1}{2}(b/a)\xi^{n+1}$$

Since this should be true for any n, then:

$$\gamma^n = -\tfrac{1}{2}(b/a)\xi^n$$

and:

$$\frac{\gamma^{n+1}}{\gamma^n} = \frac{\xi^{n+1}}{\xi^n} < 1 \tag{6-57}$$

Thus both ratios $|\gamma^{n+1}/\gamma^n|$ and $|\xi^{n+1}/\xi^n|$ are equal and less than one, indicating that the method is stable.

Solution by alternating-direction iteration

We consider now the solution of the simultaneous difference equations by an alternating-direction *iteration*. As pointed out above, when discussing simultaneous procedures, systems (6-7)–(6-8) and (6-11)–(6-12) are exactly equivalent. However, it turns out that system (6-7)–(6-8) is much more convenient for setting up the alternating-direction iteration procedure. The implicit analog of eqs. (6-7)–(6-8) is:

$$\Delta(LN_n \Delta PN_{n+1}) + QN_{ijn} = (\phi S'/\Delta t)_{ij}(PN_{i,j,n+1} - PN_{ijn} - PW_{i,j,n+1} + PW_{ijn}) \tag{6-58}$$

$$\Delta(LW_n \Delta PW_{n+1}) + QW_{ijn}$$
$$= (\phi S'/\Delta t)_{ij}(PW_{i,j,n+1} - PW_{ijn} - PN_{i,j,n+1} + PN_{ijn}) \tag{6-59}$$

The alternating-direction iteration procedure for solving this system of equations is, for the horizontal sweep:

$$\Delta_x(LN_n \Delta_x PN_{n+1}^{k+1}) + \Delta_y(LN_n \Delta_y PN_{n+1}^{k}) + QN_{ijn} = (\phi S'/\Delta t)_{ij}^k (PN_{i,j,n+1}^{k+1}$$
$$- PN_{ijn} - PW_{i,j,n+1}^{k+1} + PW_{ijn}) + SSN_{ijn}\beta_k(PN_{i,j,n+1}^{k+1} - PN_{i,j,n+1}^{k}) \tag{6-60}$$

$$\Delta_x(LW_n \Delta_x PW_{n+1}^{k+1}) + \Delta_y(LW_n \Delta_y PW_{n+1}^{k}) + QW_{ijn} = (\phi S'/\Delta t)_{ij}^k (PW_{i,j,n+1}^{k+1}$$
$$- PW_{ijn} - PN_{i,j,n+1}^{k+1} + PN_{ijn}) + SSW_{ijn}\beta_k(PW_{i,j,n+1}^{k+1} - PW_{i,j,n+1}^{k}) \tag{6-61}$$

and is, for the vertical sweep:

$$\Delta_x(LN_n \Delta_x PN_{n+1}^{k+1}) + \Delta_y(LN_n \Delta_y PN_{n+1}^{k+2}) + QN_{ijn} = (\phi S'/\Delta t)_{ij}^k (PN_{i,j,n+1}^{k+2}$$
$$- PN_{ijn} - PW_{i,j,n+1}^{k+1} + PW_{ijn}) + SSN_{ijn}\beta_k(PN_{i,j,n+1}^{k+2} - PN_{i,j,n+1}^{k+1}) \tag{6-62}$$

$$\Delta_x(LW_n \Delta_x PW_{n+1}^{k+1}) + \Delta_y(LW_n \Delta_y PW_{n+1}^{k+2}) + QW_{ijn} = (\phi S'/\Delta t)_{ij}^k (PW_{i,j,n+1}^{k+2}$$
$$- PW_{ijn} - PN_{i,j,n+1}^{k+2} + PN_{ijn}) + SSW_{ijn}\beta_k(PW_{i,j,n+1}^{k+2} - PW_{i,j,n+1}^{k+1}) \tag{6-63}$$

By working with system (6-58)–(6-59) rather than system (6-50)–(6-51), we have the situation where only one of the dependent variables is involved in the approximations to the spatial derivatives (flow terms) in each equation. Hence the last term in each of eqs. (6-60)–(6-63), involving the iteration parameter β_k, needs to include the change over the iteration of only that one variable. Similarly, the normalizing factors, SSN and SSW, need involve quantities having to do with only one phase. Let us define the following transmissibility terms:

$$ANX_{i+\frac{1}{2},j,m} = LNX_{i+\frac{1}{2},j,m} \alpha \Delta y_j/(x_{i+1} - x_i) \qquad (6\text{-}64)$$

$$ANY_{i,j+\frac{1}{2},m} = LNY_{i,j+\frac{1}{2},m} \alpha \Delta x_i/(y_{j+1} - y_j) \qquad (6\text{-}65)$$

Make similar definitions for AWX and AWY. Then, from examination of eqs. (6-17) and (6-18), we see that SSN and SSW should be calculated by:

$$SSN_{ijm} = \overline{SSN}_{ijm}/(\alpha \Delta x_i \Delta y_j) \qquad (6\text{-}66)$$

$$SSW_{ijm} = \overline{SSW}_{ijm}/(\alpha \Delta x_i \Delta y_j) \qquad (6\text{-}67)$$

where:

$$\overline{SSN}_{ijm} = ANX_{i-\frac{1}{2},j,m} + ANX_{i+\frac{1}{2},j,m} + ANY_{i,j-\frac{1}{2},m} + ANY_{i,j+\frac{1}{2},m} \qquad (6\text{-}68)$$

$$\overline{SSW}_{ijm} = AWX_{i-\frac{1}{2},j,m} + AWX_{i+\frac{1}{2},j,m} + AWY_{i,j-\frac{1}{2},m} + AWY_{i,j+\frac{1}{2},m} \qquad (6\text{-}69)$$

Next, we note the nature of the simultaneous equations that must be solved on each line. Consider one line of the horizontal sweep. If we multiply eqs. (6-60)–(6-61) through by $\alpha \Delta x_i \Delta y_j$, the following pair of equations arises:

$$-ANX_{i-\frac{1}{2},j,n}(PN^{k+1}_{i-1,j,n+1} - PN^{k+1}_{i,j,n+1}) + ANX_{i+\frac{1}{2},j,n}(PN^{k+1}_{i,j,n+1} - PN^{k+1}_{i+1,j,n+1})$$
$$+ (\phi S'/\Delta t)^k_{ij} \alpha \Delta x_i \Delta y_j (PN^{k+1}_{i,j,n+1} - PW^{k+1}_{i,j,n+1}) + \overline{SSN}_{ijn} \beta_k PN^{k+1}_{i,j,n+1}$$
$$= \overline{QN}_{ijn} + ANY_{i,j-\frac{1}{2},n}(PN^k_{i,j-1,n+1} - PN^k_{i,j,n+1}) - ANY_{i,j+\frac{1}{2},n}$$
$$\times (PN^k_{i,j,n+1} - PN^k_{i,j+1,n+1}) + (\phi S'/\Delta t)^k_{ij} \alpha \Delta x_i \Delta y_j (PN_{ijn} - PW_{ijn})$$
$$+ \overline{SSN}_{ijn} \beta_k PN^k_{i,j,n+1} \qquad (6\text{-}70)$$

$$-AWX_{i-\frac{1}{2},j,n}(PW^{k+1}_{i-1,j,n+1} - PW^{k+1}_{i,j,n+1}) + AWX_{i+\frac{1}{2},j,n}(PW^{k+1}_{i,j,n+1} - PW^{k+1}_{i+1,j,n+1})$$
$$+ (\phi S'/\Delta t)^k_{ij} \alpha \Delta x_i \Delta y_j (PW^{k+1}_{i,j,n+1} - PN^{k+1}_{i,j,n+1}) + \overline{SSW}_{ijn} \beta_k PW^{k+1}_{i,j,n+1}$$
$$= \overline{QW}_{ijn} + AWY_{i,j-\frac{1}{2},n}(PW^k_{i,j-1,n+1} - PW^k_{i,j,n+1}) - AWY_{i,j+\frac{1}{2},n}$$
$$\times (PW^k_{i,j,n+1} - PW^k_{i,j+1,n+1}) + (\phi S'/\Delta t)^k_{ij} \alpha \Delta x_i \Delta y_j (PW_{ijn} - PN_{ijn})$$
$$+ \overline{SSW}_{ijn} \beta_k PW^k_{i,j,n+1} \qquad (6\text{-}71)$$

If we arrange the unknowns in the order of $PN_{1,j}$, $PW_{1,j}$, $PN_{2,j}$, $PW_{2,j}$, $PN_{3,j}$, etc. (i.e., interleave the unknowns PN and PW), then the coefficient matrix takes on the structure shown in Fig. 30. This matrix can be partitioned into a tridiagonal array of 2×2 submatrices; it is referred to as a bitridiagonal system of equations. While a general bitridiagonal system has all the 2×2 submatrices full, this is a special case in that the off-diagonal submatrices are diagonal. It can be seen the coefficient matrix is a band matrix with a

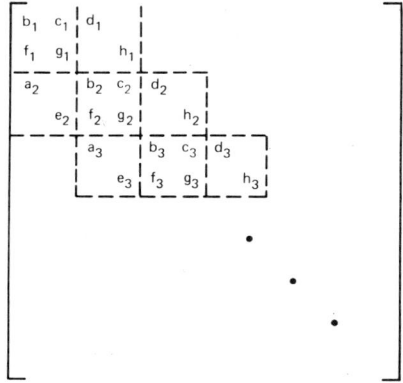

Fig. 30. Form of coefficient matrix for one line of alternating-direction iteration in simultaneous solution for two phase pressures.

width, W, of 5. Hence the band algorithm developed in Chapter 5 can be used to solve these equations.

Finally, we remark on the choice of iteration parameters, β_k. It has been our experience that if a problem is solved two ways, the first being simultaneous solution with A.D.I. and the second being another procedure using an A.D.I. solution of the pressure equation (where there is only one unknown, such as PA), the same parameters that work for solving the pressure equation carry over and serve just as well for the simultaneous A.D.I. However, the coefficient matrix of the pressure equation is determined solely by the distribution of LTX and LTY; these involve the sums of the individual phase transmissibilities — i.e. $(ANX + AWX)$ and $(ANY + AWY)$. Hence we assert that the distribution of the quantities $(ANX + AWX)$ and $(ANY + AWY)$ determine the choice of parameters for simultaneous A.D.I.

Solution by strongly implicit procedure

Weinstein et al. (1970) extended the strongly implicit procedure of Stone (1968) to the solution of two and three simultaneous equations at each grid point. As mentioned in Chapter 5, S.I.P. has also been extended to the solution of three-dimensional problems.

In extending S.I.P. to solve two- or three-phase flow problems, it is merely necessary to replace the scalar elements, P_{ij} and Q_{ij}, by two- or three-component vectors, and to replace the scalar elements B_{ij}, D_{ij}, etc., of eq. (5-151) by 2×2 or 3×3 matrices. The elements of the approximate factors, such as b_{ij}, c_{ij}, etc., are also 2×2 or 3×3 matrices. In a manner exactly parallel to the development of the S.I.P. algorithm in Chapter 5, but replacing the multiplication and division operations by 2×2 or 3×3 matrix multiplications or inversions, Weinstein et al. developed the algorithms for the two-phase and three-phase iterations. Tests have shown that the choice of iteration parameters follows the same considerations as for the single-phase case. Many problems that have been difficult (or impossible) to solve by alternating-direction iteration have been found to be much easier to solve by the use of S.I.P.

Direct solution

The direct solution of the simultaneous difference equations (6-58)–(6-59) has also been implemented (by the author). As with alternating-direction iteration, the unknowns are interleaved; they are put in the order $PN_{1,1}$, $PW_{1,1}$, $PN_{2,1}$, $PW_{2,1}$, ..., $PN_{I,1}$, $PW_{I,1}$, $PN_{1,2}$, $PW_{1,2}$, ..., $PN_{I,2}$, $PW_{I,2}$, ..., $PN_{1,J}$, $PW_{1,J}$, ..., $PN_{I,J}$, $PW_{I,J}$. The coefficients have a band structure with a width, W, equal to $4I+1$. For known S'_{ij}, this is a linear set of equations, which can be solved using the band algorithm of Chapter 5.

Problems having straight-line capillary pressure curves (constant S') lend themselves nicely to direct solution. The computing labor goes up as $I^3 J$, so the direct solution is most applicable to long, narrow grids, such as are frequently used for cross-section studies. As a rough estimate, it appears that for $I = 15$ to 20, the direct solution takes as much computing time as 10 to 15 alternating-direction iterations.

Straight-line capillary pressure curves are rare, however. For the usual case of variable S', it appears that a good strategy is, for each time step, to use some iterative method to get a reasonably good estimate of $S_{i,j,n+1}$, use this to compute $S'_{i,j,n+\frac{1}{2}}$ (as shown in the next section), and then to finish the simultaneous solution for the pressures by use of the direct solution algorithm. This combination approach has been found useful in a number of cross-section studies.

Calculation of the nonlinear coefficient, S'

Up to now, mobilities have been assigned the subscript n to indicate that they are evaluated as functions of SN_{ijn}. No such assignment was made for S'_{ij}, in order to leave open the question of how it should be evaluated. Since the time derivatives have been approximated by backward differences, these difference equations are only first-order correct in time. Hence, one is tempted to evaluate S'_{ij} at t_n. Results of such calculations have large time truncation errors that can be sufficiently reduced only by using excessively small time steps. This time truncation error manifests itself particularly in the calculated material balance (which is discussed more fully, below). Thus, even though the use of S'_{ijn} brings about an error of the same order, $\mathcal{O}(\Delta t)$, as that caused by the use of backward-difference equations, it is clear that a better procedure for evaluating S'_{ij} is needed.

The use of iteration for calculating the phase pressures allows reevaluations of S'_{ij} at the end of each iteration as:

$$(S')^k_{i,j,n+\frac{1}{2}} = \frac{SN^k_{i,j,n+1} - SN_{ijn}}{PC^k_{i,j,n+1} - PC_{ijn}} = \frac{S_n(PC^k_{i,j,n+1}) - SN_{ijn}}{(PN-PW)^k_{i,j,n+1} - (PN-PW)_{ijn}} \quad (6\text{-}72)$$

where SN_{ijn} is the finite-difference solution for the nonwetting phase saturation S_n (and is evaluated from PC_{ijn}). Equation (6-72) is the most straightforward way to reevaluate S'; some programs use a damped (or "underrelaxed") version of (6-72). For the most part, this method has worked well and does not appreciably slow down the iteration for the phase pressures.

After convergence of the iteration, we have the result:

$$S'_{ij} = \frac{SN_{i,j,n+1} - SN_{ijn}}{PC_{i,j,n+1} - PC_{ijn}} \qquad (6\text{-}73)$$

Substitution into eqs. (6-58)–(6-59) gives:

$$\Delta(LN_n \Delta PN_{n+1}) + QN_{ijn} = (\phi_{ij}/\Delta t)(SN_{i,j,n+1} - SN_{ijn}) \qquad (6\text{-}74)$$

$$\Delta(LW_n \Delta PW_{n+1}) + QW_{ijn} = (\phi_{ij}/\Delta t)(SW_{i,j,n+1} - SW_{ijn}) \qquad (6\text{-}75)$$

Material balance

The material balance equation states that the total rate of input of either phase should equal the total rate of accumulation of that phase. In a finite-difference form, the material balance equation for the nonwetting phase is:

$$\sum_i \sum_j \overline{QN}_{ijn} = \sum_i \sum_j \phi_{ij} \alpha \Delta x_i \Delta y_j (SN_{i,j,n+1} - SN_{ijn})/\Delta t \qquad (6\text{-}76)$$

If we multiply eq. (6-74) through by $\alpha \Delta x_i \Delta y_j$ and sum over all grid points, then we get:

$$\sum_i \sum_j \alpha \Delta x_i \Delta y_j \Delta(LN_n \Delta PN_{n+1}) + \sum_i \sum_j \overline{QN}_{ijn}$$

$$= \sum_i \sum_j (\phi_{ij}/\Delta t) \alpha \Delta x_i \Delta y_j (SN_{i,j,n+1} - SN_{ijn}) \qquad (6\text{-}77)$$

Upon examination of the first summation of eq. (6-77), we note that it is the sum of flow terms such as:

$$ANX_{i+\frac{1}{2},j,n}(PN_{i+1,j,n+1} - PN_{i,j,n+1})$$

coming from eq. (6-74) written for one grid point, which exactly cancels another flow term:

$$ANX_{i-\frac{1}{2},j,n}(PN_{i,j,n+1} - PN_{i-1,j,n+1})$$

coming from eq. (6-74) written for a neighboring grid point. All the flow terms cancel out, so that:

$$\sum_i \sum_j \alpha \Delta x_i \Delta y_j \Delta(LN_n \Delta PN_{n+1}) = 0$$

Thus, eq. (6-77) reduces to the material balance equation (6-76). This identity would not be true (and hence the material balance would not be satisfied) without the iteration on S'_{ij}.

Because the material balance is an identity, it is used as a check on the calculations as well as a check on the convergence of the iteration. Note, however, that the material balance is not (in this case, at least) a check on the truncation error of the solution.

Summary

A variety of simulators based on the simultaneous solution for two dependent variables in two-phase flow problems, and for three dependent variables in three-phase flow problems, has been described in the literature. Following the original work of Douglas, Peaceman and Rachford (1959), pioneering work was presented by Coats et al. (1967), who extended the simultaneous solution method to three dimensions, and by Peery and Herron (1969) and Sheffield (1969) who were among the first to extend it to three-phase flow problems. Other noteworthy examples of the simultaneous solution method appearing in the literature are: Snyder (1969); Blair and Weinaug (1969); the last of the three methods described by MacDonald and Coats (1970), Letkeman and Ridings (1970); Sonier et al. (1973); Settari and Aziz (1974); and Trimble and McDonald (1976).

The large number of papers describing simulators based on the simultaneous solution method attest to a popularity that is due to its reliability, computational stability, and frequently better quality of answers, compared with some of the sequential methods we shall describe below. These benefits are considered sufficient to overcome the greater cost resulting from the increased computational labor that is inherent in the simultaneous solution method.

SEQUENTIAL NUMERICAL SOLUTION

Introduction

We have noted earlier that the simultaneous differential equations in phase pressures, eqs (6-7)—(6-8), can be replaced by another pair of differential equations. The first equation of the alternative pair is a pressure equation, such as (6-11); the second equation may involve capillary pressure, as does eq. (6-12), or it may involve saturation, as do eqs. (6-1), (6-2), or (6-16). This suggests that we consider solving the numerical analogs of two differential equations in sequence each time step. The advantage is obvious, in that only one unknown at each grid point needs to be solved for in each partial step. The difference equations should be much easier to solve, and a substantial reduction in the computing labor should result. Furthermore, additional flexibility is available in that the same numerical method need not be used for both differential equations; rather, the most appropriate method can be chosen for each one.

Leapfrog method

In addition to the simultaneous solution method discussed above, Douglas, Peaceman and Rachford (1959) proposed the first sequential method, which they termed the "leapfrog" method. It is based on eqs. (6-11)—(6-12), and consists of a relatively minor change in its difference analog, eqs. (6-50)—(6-51). That change is to replace PC_{n+1} in eq. (6-50) by PC_n:

$$\Delta(LT_n \Delta PA_{n+1}) + \tfrac{1}{2}\Delta(LD_n \Delta PC_n) + (QN + QW)_{ijn} = 0 \tag{6-78}$$

$$\tfrac{1}{2}\Delta(LT_n \Delta PC_{n+1}) + \Delta(LD_n \Delta PA_{n+1}) + (QN - QW)_{ijn}$$
$$= 2(\phi S'/\Delta t)_{ij}(PC_{i,j,n+1} - PC_{ijn}) \tag{6-79}$$

Equation (6-78) is first solved for $PA_{i,j,n+1}$ using one of the methods of Chapter 5 for solving elliptic problems in one dependent variable. These values of $PA_{i,j,n+1}$ can then be substituted into eq. (6-79); it, in turn, needs to be solved for only one variable at each grid point, namely $PC_{i,j,n+1}$.

While the leapfrog method of Douglas, Peaceman and Rachford (1959) apparently has not been widely used, it does share many features of other sequential methods which have become more popular. It is one of the easiest of the sequential methods to analyze for stability, so we shall use it as a prototype for studying stability and also to examine a phenomenon known as saturation creep.

Stability

To examine the stability of the leapfrog equations (6-78)—(6-79), we proceed in exactly the same manner that led to eqs. (6-54)—(6-55), obtaining now:

$$-(X_p + Y_q)(a\gamma^{n+1} + \tfrac{1}{2}b\xi^n) = 0 \tag{6-80}$$

$$-(X_p + Y_q)(b\gamma^{n+1} + \tfrac{1}{2}a\xi^{n+1}) = \xi^{n+1} - \xi^n \tag{6-81}$$

From eq. (6-80) we obtain:

$$\gamma^{n+1} = -\tfrac{1}{2}(b/a)\xi^n \tag{6-82}$$

Substitution of eq. (6-82) into (6-81) gives:

$$-\tfrac{1}{2}(X_p + Y_q)[a\xi^{n+1} - (b^2/a)\xi^n] = \xi^{n+1} - \xi^n$$

or:

$$\frac{\xi^{n+1}}{\xi^n} = \frac{1 + \tfrac{1}{2}(b^2/a)(X_p + Y_q)}{1 + \tfrac{1}{2}a(X_p + Y_q)} \tag{6-83}$$

Since $a > 0$ and $|b| < a$, (b^2/a) is less than a, so this ratio is less than one. From eq. (6-82), we can also write:

$$\gamma^{n+2} = -\tfrac{1}{2}(b/a)\xi^{n+1}$$

or:

$$\gamma^{n+2}/\gamma^{n+1} = \xi^{n+1}/\xi^n$$

Hence we can conclude also that:

$$|\gamma^{n+1}/\gamma^n| < 1 \tag{6-84}$$

Therefore the leapfrog method is stable.

Saturation creep

The leapfrog method has given satisfactory results in a limited number of calculations, although the time truncation error tends to be somewhat larger than that obtained with simultaneous solution; this, of course, is not surprising.

The leapfrog method does give rise to an anomalous situation where nontrivial changes in saturation take place in regions of the reservoir where there is no injection but where one of the relative permeabilities is zero. This change, which is usually not large but does persist, is known as saturation creep. It is not a physically real phenomenon, of course, and does not occur in simulators using the simultaneous solution method.

To understand the origin of saturation creep, we note that if the wetting-phase relative permeability is zero (as may be the case in the uninvaded oil zone), then, by eqs. (6-25) and (6-26):

$$LTX = LDX = LNX$$
$$LTY = LDY = LNY$$

Upon subtracting eq. (6-78) from (6-79) and taking $QN = QW = 0$, we get:

$$\tfrac{1}{2}\Delta(LN_n \Delta PC_{n+1}) - \tfrac{1}{2}\Delta(LN_n \Delta PC_n) = 2(\phi S'/\Delta t)_{ij}(PC_{i,j,n+1} - PC_{ijn})$$

or:

$$\tfrac{1}{2}\Delta(LN_n \Delta[PC_{n+1} - PC_n]) = 2(\phi_{ij}/\Delta t)(SN_{i,j,n+1} - SN_{ijn}) \tag{6-85}$$

On the other hand, if the nonwetting-phase relative permeability is zero (as may be the case in the aquifer), then, by eqs. (6-25) and (6-26):

$$LTX = -LDX = LWX$$
$$LTY = -LDY = LWY$$

Adding equations (6-78) and (6-79) and taking $QN = QW = 0$ yields:

$$\tfrac{1}{2}\Delta(LW_n \Delta PC_{n+1}) - \tfrac{1}{2}\Delta(LW_n \Delta PC_n) = 2(\phi S'/\Delta t)_{ij}(PC_{i,j,n+1} - PC_{ijn})$$

or:

$$\tfrac{1}{2}\Delta(LW_n\Delta[PC_{n+1} - PC_n]) = 2(\phi_{ij}/\Delta t)(SN_{i,j,n+1} - SN_{ijn}) \qquad (6\text{-}86)$$

If the capillary pressure curve is steep, small fluctuations in saturation caused by round-off error or by incomplete iterations can induce relatively large fluctuations in *PC*. From eq. (6-85) or (6-86), we see that these spatial fluctuations in *PC* will induce changes in saturation from one time step to the next.

Let us reconsider the simultaneous solution method. For that method, if the wetting-phase relative permeability is zero, we get, instead of eq. (6-85):

$$\tfrac{1}{2}\Delta(LN_n\Delta[PC_{n+1} - PC_{n+1}]) = 2(\phi_{ij}/\Delta t)(SN_{i,j,n+1} - SN_{ijn})$$

while if the nonwetting-phase relative permeability is zero, we get, instead of eq. (6-86):

$$\tfrac{1}{2}\Delta(LW_n\Delta[PC_{n+1} - PC_{n+1}]) = 2(\phi_{ij}/\Delta t)(SN_{i,j,n+1} - SN_{ijn})$$

In either case:

$$SN_{i,j,n+1} = SN_{ijn}$$

so that saturation creep cannot occur if simultaneous solution is being used.

Other sequential solution methods

A more popular choice for a sequential solution procedure than the leapfrog method (which involves solving first for *PA* and then for *PC*) has been to solve first for one of the phase pressures (usually the oil pressure) and then to solve for saturation, either explicitly or implicitly. Examples in the literature are: Fagin and Stewart (1966); Breitenbach et al. (1968); and the first two methods described by MacDonald and Coats (1970).

The differential equation for pressure (6-11) is modified so that the dependent variables are one phase pressure and the capillary pressure. Assuming oil is the nonwetting phase, we choose the pressure of that phase to be a dependent variable. Then we write:

$$p_{\text{avg}} = (p_n + p_w)/2 = p_n - \tfrac{1}{2}p_c \qquad (6\text{-}87)$$

Substitution of eq. (6-87) into the pressure equation (6-11) yields:

$$\nabla \cdot (\lambda_t \nabla p_n) + \tfrac{1}{2}\nabla \cdot [(\lambda_d - \lambda_t)\nabla p_c] + Q_n + Q_w = 0$$

But:

$$\lambda_d - \lambda_t = -2\lambda_w$$

so that the pressure equation becomes:

$$\nabla \cdot (\lambda_t \nabla p_n) - \nabla \cdot (\lambda_w \nabla p_c) + Q_n + Q_w = 0 \qquad (6\text{-}88)$$

The difference analog is:

$$\Delta(LT_n \Delta PN_{n+1}) - \Delta(LW_n \Delta PC_n) + (QN + QW)_{ijn} = 0 \quad (6\text{-}89)$$

which can be solved for $PN_{i,j,n+1}$ implicitly by one of the methods of Chapter 5 for solving elliptic equations in one dependent variable.

Since the nonwetting phase pressure at the new time level, t_{n+1}, is now known, the difference analog for eq. (6-8):

$$\Delta(LW_n \Delta[PN_{n+1} - PC_{n+1}]) + QW_{ijn} = -(\phi S'/\Delta t)_{ij}(PC_{i,j,n+1} - PC_{ijn}) \quad (6\text{-}90)$$

can be solved implicitly for the new capillary pressure at each grid point.

This method is also subject to saturation creep when the nonwetting-phase relative permeability is zero (as in an aquifer region). In this situation, we have:

$$LTX = LWX \quad \text{and} \quad LTY = LWY$$

so that subtraction of eq. (6-90) from (6-89), taking $QN = QW = 0$, gives:

$$\Delta(LW_n \Delta[PC_{n+1} - PC_n]) = (\phi_{ij}/\Delta t)(SN_{i,j,n+1} - SN_{ijn})$$

Again we see that if the capillary pressure curve is steep, small spatial fluctuations in saturation lead to larger spatial fluctuations in PC; these larger fluctuations in turn induce physically unreal saturation changes from one time step to the next.

Other variations on this sequential method are possible. For example, the wetting-phase pressure can be the unknown in the pressure equation, and the difference analog of eq. (6-7) can be used to calculate the new capillary pressure. In all cases, the significant aspect of the method is that the capillary pressure term is evaluated explicitly in the pressure equation. Combined with implicit evaluation of capillary pressure in the saturation equation, this turns out to be the source of the anomalous saturation creep behavior.

Sequential solution method using total velocity

Spillette et al. (1973) proposed a sequential simulator based on the use of eq. (6-88) for the pressure equation and a difference analog of the Buckley-Leverett form of the saturation equation, eq. (6-16). The significant innovation is that the total velocity is calculated after solution of the pressure equation, and this velocity is then used for the saturation equation. We shall demonstrate two basic properties of this method, which contribute to its advantage over other sequential methods. First, the total velocity calculated from the pressures (after solution of the pressure equation) is the same whether the capillary pressure term in the pressure equation is evaluated explicitly or implicitly. Second, when either relative permeability is zero, there is no anomalous saturation creep.

Instead of using eq. (6-89) as the difference analog of the pressure differential equation (6-88), let us write the following analog:

$$\Delta(LT_n\Delta PN_{n+1}) - \Delta(LW_n\Delta PC_m) + (QN + QW)_{ijn} = 0 \tag{6-91}$$

The subscript m is used on the capillary pressure term in order that we do not commit ourselves, at this point, as to whether it is evaluated explicitly or implicitly. Now, for each grid interval, the difference analogs for the components of the phase velocities are by eqs. (6-13) and (6-14):

$$VNX_{i+\frac{1}{2},j} = -LNX_{i+\frac{1}{2},j,n}(PN_{i+1,j,n+1} - PN_{i,j,n+1})/(x_{i+1} - x_i)$$

$$VNY_{i,j+\frac{1}{2}} = -LNY_{i,j+\frac{1}{2},n}(PN_{i,j+1,n+1} - PN_{i,j,n+1})/(y_{j+1} - y_j)$$

Similar expressions hold for $VWX_{i+\frac{1}{2},j}$ and $VWY_{i,j+\frac{1}{2}}$. Comparison with eqs. (6-17), (6-18), and (6-21) gives:

$$\Delta(LN_n\Delta PN_{n+1}) = -\frac{(VNX_{i+\frac{1}{2},j} - VNX_{i-\frac{1}{2},j})}{\Delta x_i} - \frac{(VNY_{i,j+\frac{1}{2}} - VNY_{i,j-\frac{1}{2}})}{\Delta y_j} \tag{6-92}$$

We recognize the right-hand side of eq. (6-92) to be the discrete analog of $-\nabla \cdot \vec{v}_n$. Therefore, let us designate it by $-\Delta(VN)$, and rewrite eq. (6-92) as:

$$\Delta(LN_n\Delta PN_{n+1}) = -\Delta(VN)$$

Now, while $PW_{i,j,n+1}$ has not been solved for, we can obtain it from the solution for the nonwetting-phase pressure by using the equation:

$$PW_{i,j,n+1} = PN_{i,j,n+1} - PC_{i,j,m}$$

Then we can write the following expression for the wetting-phase velocity:

$$\Delta(VW) = -\Delta(LW_n\Delta PW_{n+1}) = -\Delta(LW_n\Delta PN_{n+1}) + \Delta(LW_n\Delta PC_m)$$

The total velocity for each grid interval is:

$$VTX_{i+\frac{1}{2},j} = VNX_{i+\frac{1}{2},j} + VWX_{i+\frac{1}{2},j}$$

$$VTY_{i,j+\frac{1}{2}} = VNY_{i,j+\frac{1}{2}} + VWY_{i,j+\frac{1}{2}}$$

so that:

$$\Delta(VT) = \Delta(VN) + \Delta(VW)$$

$$\Delta(VT) = -\Delta(LN_n\Delta PN_{n+1}) - \Delta(LW_n\Delta PN_{n+1}) + \Delta(LW_n\Delta PC_m)$$

$$\Delta(VT) = -\Delta(LT_n\Delta PN_{n+1}) + \Delta(LW_n\Delta PC_m) \tag{6-93}$$

Substitution of eq. (6-93) into (6-91) gives:

$$\Delta(VT) = (QW + QN)_{ijn} \tag{6-94}$$

which we recognize to be the difference analog of eq. (1-53). Equation (6-94) shows that the values obtained for total velocity are independent of m; we have therefore demonstrated the proposition that the total velocity is insensitive to the time level at which capillary pressure is evaluated in the pressure equation.

While Spillette et al. chose to solve for oil pressure in the pressure equation, other choices are possible. In particular, one might solve for the average pressure, using eq. (6-78). In that case, we would define:

$$PN_{i,j,n+1} = PA_{i,j,n+1} + \tfrac{1}{2}PC_{i,j,m}$$
$$PW_{i,j,n+1} = PA_{i,j,n+1} - \tfrac{1}{2}PC_{i,j,m}$$

and arrive at the same result for the total velocity, namely eq. (6-94), thus showing VTX and VTY again to be independent of the subscript on the capillary pressure term in (6-78).

Let us now consider the difference analog for the saturation equation (6-16). First, we define interval values for the quantities f_w and $f_w \lambda_n$:

$$FWX_{i+\frac{1}{2},j,m} = LWX_{i+\frac{1}{2},j,m}/(LNX_{i+\frac{1}{2},j,m} + LWX_{i+\frac{1}{2},j,m}) \qquad (6\text{-}95)$$
$$FWY_{i,j+\frac{1}{2},m} = LWY_{i,j+\frac{1}{2},m}/(LNY_{i,j+\frac{1}{2},m} + LWY_{i,j+\frac{1}{2},m}) \qquad (6\text{-}96)$$
$$FWLNX_{i+\frac{1}{2},j,m} = FWX_{i+\frac{1}{2},j,m} LNX_{i+\frac{1}{2},j,m} \qquad (6\text{-}97)$$
$$FWLNY_{i,j+\frac{1}{2},m} = FWY_{i,j+\frac{1}{2},m} LNY_{i,j+\frac{1}{2},m} \qquad (6\text{-}98)$$

Expressions analogous to (6-17) and (6-18) are used to define:

$$\Delta_x(FWLN_m \Delta_x PC_n) \quad \text{and} \quad \Delta_y(FWLN_m \Delta_y PC_n)$$

which may then be added together to form the two-dimensional operator:

$$\Delta(FWLN_m \Delta PC_n) = \Delta_x(FWLN_m \Delta_x PC_n) + \Delta_y(FWLN_m \Delta_y PC_n) \qquad (6\text{-}99)$$

For the discrete analog of the convection term, we have the operators:

$$\Delta_x(FW_m VT) = (FWX_{i+\frac{1}{2},j,m} VTX_{i+\frac{1}{2},j} - FWX_{i-\frac{1}{2},j,m} VTX_{i-\frac{1}{2},j})/\Delta x_i$$
$$\Delta_y(FW_m VT) = (FWY_{i,j+\frac{1}{2},m} VTY_{i,j+\frac{1}{2}} - FWY_{i,j-\frac{1}{2},m} VTY_{i,j-\frac{1}{2}})/\Delta y_j$$

and the combined two-dimensional operator:

$$\Delta(FW_m VT) = \Delta_x(FW_m VT) + \Delta_y(FW_m VT) \qquad (6\text{-}100)$$

which is the analog of $\nabla \cdot (f_w \vec{v}_t)$.

The difference analog of the saturation equation (6-16) is then:

$$-\Delta(FWLN_n PC_m) - \Delta(FW_n VT) + QW_{ijn} = (\phi_{ij}/\Delta t)(SW_{i,j,n+1} - SW_{ijn}) \qquad (6\text{-}101)$$

Spillette et al. point out that an explicit treatment of the capillary pressure term (i.e., $m = n$) should be used only for easy-to-solve problems; they recommend that in most cases a semi-implicit treatment of the capillary pressure term should be used, wherein:

$$PC_{i,j,m} = PC_{ijn} + (dp_c/dS_w)_{ijn}(SW_{i,j,n+1} - SW_{ijn}) \qquad (6\text{-}102)$$

(The topic of semi-implicit treatment will be discussed more fully in the next section.) The substitution of eq. (6-102) into (6-101) produces a linear set of equations which, because of the spatial differencing, is implicit in $SW_{i,j,n+1}$. Spillette et al. provide several options for the solution of the saturation equation; one is to treat it as a parabolic equation and take one double sweep of an alternating-direction *implicit* procedure; other options are to iterate using S.I.P., or to use direct solution.

Examination of eq. (6-101) shows why this method is not subject to saturation creep when either relative permeability is zero. By eqs. (6—95)—(6-98), we have:

$$FWLNX_{i+\frac{1}{2},j,m} = LWX_{i+\frac{1}{2},j,m} LNX_{i+\frac{1}{2},j,m} / (LNX_{i+\frac{1}{2},j,m} + LWX_{i+\frac{1}{2},j,m})$$

and:

$$FWLNY_{i,j+\frac{1}{2},m} = LWY_{i,j+\frac{1}{2},m} LNY_{i,j+\frac{1}{2},m} / (LNY_{i,j+\frac{1}{2},m} + LWY_{i,j+\frac{1}{2},m})$$

which are zero if *either* relative permeability is zero. Since these coefficients multiply the capillary pressure differences in eq. (6-101), the mechanism for producing saturation creep is absent.

IMPLICIT AND SEMI-IMPLICIT MOBILITIES

Limitations resulting from use of explicit mobilities

Up to now, all mobilities have been specified as being evaluated as functions of saturation at the old time level, t_n. This is referred to as the use of *explicit* mobilities. Simulators based on explicit mobilities are satisfactory for a great variety of field problems, but there are two particular types of problems where they fail badly. One is the coning problem, which involves converging radial flow into a well with small radial increments near the boundary that represents the well. The second is the gas percolation problem, in which gas comes out of solution and rises up through the reservoir. Both of these types of problems involve high velocity in one or both phases. In the coning problem, the high velocity is a result of the converging flow. In the gas percolation problem, high gas velocity is a result of the low gas viscosity and the large density difference between oil and gas. In both cases, unless the time steps are cut to impractically small values, very large oscillations in the calculated saturations occur in the high-velocity regions.

The reason that the stability analyses given earlier in this chapter do not predict this instability (or a severe time-step restriction for stability) is that the mobility was assumed to be constant; this, of course, is not the case. It is not hard to perform a stability analysis which does not assume constant mobility, but then it is necessary or desirable to make some other simplifications.

Perhaps the easiest approach is to start with eq. (6-101). Since high-velocity problems are characterized by the dominance of the convection term over the capillary pressure term, we shall assume $p_c = 0$. In addition, let us drop QW. Then eq. (6-101) simplifies to:

$$-\Delta(FW_n VT) = (\phi_{ij}/\Delta t)(SW_{i,j,n+1} - SW_{ijn}) \tag{6-103}$$

If we further simplify the problem to one dimension and assume the total velocity is constant, then eq. (6-103) becomes:

$$-VTX \frac{FWX_{i+\frac{1}{2},n} - FWX_{i-\frac{1}{2},n}}{\Delta x_i} = \phi_i \frac{SW_{i,n+1} - SW_{in}}{\Delta t} \tag{6-104}$$

Assume VTX is positive. Then we see that, except for the additional factor of porosity, ϕ_i, eq. (6-104) is identical to (4-7), provided the time-weighting parameter, θ, is taken equal to zero in eq. (4-7). We examined the stability of this equation in Chapter 4 and found that it is unconditionally unstable for midpoint weighting ($W = 1/2$), while for upstream weighting ($W = 1$) it is stable provided (4-17) is satisfied. If we include the additional factor of porosity, the condition for stability is then:

$$\frac{VTX(f')\Delta t}{\phi_i \Delta x_i} \leqslant 1 \tag{6-105}$$

From this condition it is easy to see why the time step must be very small, when the velocity becomes very large, in order to prevent oscillatory solutions.

The analysis could be extended to two dimensions and include nonzero capillary pressure. The effect of nonzero p_c is to relax the time step restriction somewhat, but we can still expect that the use of explicit mobilities will cause large oscillations in saturation when the flow through any grid block in one time step, i.e.:

$$\frac{VTX \cdot \Delta t}{\phi \Delta x} \quad \text{or} \quad \frac{VTY \cdot \Delta t}{\phi \Delta y}$$

is very much larger than one in magnitude.

Implicit mobilities

The discussion in Chapter 4 showed that increasing the implicitness of the nonlinear convection term will improve the stability. Equation (4-7) was shown to be unconditionally stable for either midpoint or upstream weighting ($1/2 \leqslant W \leqslant 1$) when the spatial difference was made implicit by choosing $\theta = 1$. In the light of this analysis, the most obvious way to stabilize two-phase flow difference equations is to make the mobilities implicit, that is, to evaluate them at the new time level, t_{n+1}.

Blair and Weinaug (1969) developed such a simulator, using implicit mobilities, that involved simultaneous solution for the dependent variables $PW_{i,j,n+1}$ and $PN_{i,j,n+1}$. If the mobilities are simply updated each iteration, it can be shown that the resulting iteration will not converge for any time step that violates eq. (6-105). It was necessary for Blair and Weinaug to use the more elaborate solution technique known as Newtonian iteration, which is a general method for solving a system of nonlinear equations. This may be described briefly as follows.

Suppose we have N equations in N variables:

$$f_1(P_1, P_2, \ldots, P_N) = 0$$
$$f_2(P_1, P_2, \ldots, P_N) = 0$$
$$\vdots$$
$$f_N(P_1, P_2, \ldots, P_N) = 0$$

then each iteration consists of solving the following *linear* equations for P^{k+1}:

$$(P_1^{k+1} - P_1^k)\left(\frac{\partial f_1}{\partial P_1}\right)^k + (P_2^{k+1} - P_2^k)\left(\frac{\partial f_1}{\partial P_2}\right)^k + \ldots + (P_N^{k+1} - P_N^k)\left(\frac{\partial f_1}{\partial P_N}\right)^k = -f_1^k$$

$$(P_1^{k+1} - P_1^k)\left(\frac{\partial f_2}{\partial P_1}\right)^k + (P_2^{k+1} - P_2^k)\left(\frac{\partial f_2}{\partial P_2}\right)^k + \ldots + (P_N^{k+1} - P_N^k)\left(\frac{\partial f_2}{\partial P_N}\right)^k = -f_2^k$$

$$\vdots$$

$$(P_1^{k+1} - P_1^k)\left(\frac{\partial f_N}{\partial P_1}\right)^k + (P_2^{k+1} - P_2^k)\left(\frac{\partial f_N}{\partial P_2}\right)^k + \ldots + (P_N^{k+1} - P_N^k)\left(\frac{\partial f_N}{\partial P_N}\right)^k = -f_N^k$$

These linear equations turn out to have the same structure as those obtained with explicit mobility simulators, so the same solution techniques can be used. The coefficients are somewhat different, of course, and entail a considerable amount of additional computational effort. Because Newtonian iteration will not converge without a reasonable initial estimate, a time-step limitation for convergence still exists, although the allowable time step is much larger than that required for explicit mobility simulation.

Semi-implicit mobilities

An effective substitute for implicit mobilities has been found that is even more effective at overcoming the time-step restriction, without increasing the computational effort much beyond that required for explicit mobilities. This involves estimating the mobilities (and capillary pressures) at the new time level by an extrapolation, e.g.:

$$(\lambda_n)_{i,j,n+1} = (\lambda_n)_{ijn} + (SN_{i,j,n+1} - SN_{ijn})\left(\frac{d\lambda_n}{dS_n}\right)_{ijn} \qquad (6\text{-}106)$$

$$(\lambda_w)_{i,j,n+1} = (\lambda_w)_{ijn} + (SN_{i,j,n+1} - SN_{ijn})\left(\frac{d\lambda_w}{dS_n}\right)_{ijn} \qquad (6\text{-}107)$$

These extrapolated mobilities are referred to as *semi-implicit* mobilities.

A linearized stability analysis can be applied to a semi-implicit replacement for eq. (4-7), namely:

$$-(v/\Delta x)[F_{i+\frac{1}{2},n} + (S_{i+\frac{1}{2},n+1} - S_{i+\frac{1}{2},n})(df/dS)_{i+\frac{1}{2},n}$$
$$-F_{i-\frac{1}{2},n} - (S_{i-\frac{1}{2},n+1} - S_{i-\frac{1}{2},n})(df/dS)_{i-\frac{1}{2},n}] = (S_{i,n+1} - S_{in})/\Delta t$$

If we linearize this equation by assuming df/dS is constant and equal to f', then we have:

$$-(vf'/\Delta x)[S_{i+\frac{1}{2},n+1} - S_{i-\frac{1}{2},n+1}] = (S_{i,n+1} - S_{in})/\Delta t$$

with:

$$S_{i+\frac{1}{2},n+1} = (W)S_{i,n+1} + (1-W)S_{i+1,n+1}$$

Thus the linearized form of the semi-implicit mobility equation is exactly the same as the linearized form of the implicit mobility equation, and this has already been shown to be stable for $1/2 \leq W \leq 1$.

MacDonald and Coats (1970) and Letkeman and Ridings (1970) implemented semi-implicit mobilities and capillary pressure in simulators designed to perform coning calculations. In their formulation, products of $(S_{i,j,n+1} - S_{ijn})$ and $(P_{i,j,n+1} - P_{ijn})$ were dropped in order to keep the equations more nearly linear. Nolen and Berry (1972) also implemented a semi-implicit simulator; they avoided the need to drop these product terms by using a Newtonian iteration to solve the resulting nonlinear equations. In this case, the use of Newtonian iteration did not add significantly to the calculation time. Nolen and Berry point out that large material balance errors can result for dissolved-gas systems if these product terms are neglected.

Semi-implicit mobilities are also provided as an option in the sequential simulator using total velocities described by Spillette et al. (1973). Both Nolen and Berry (1972) and Spillette et al. provide for the use of a "chord slope" for the derivatives in eqs. (6-106)—(6-107). They propose that the derivatives, $(d\lambda/dS)_{ijn}$, be replaced by:

$$\overline{\left(\frac{d\lambda_n}{dS_n}\right)}_{ij} = \frac{\lambda_n(SN_{ijn} + \delta S) - \lambda_n(SN_{ijn})}{\delta S} \quad (6\text{-}108)$$

$$\overline{\left(\frac{d\lambda_w}{dS_n}\right)}_{ij} = \frac{\lambda_w(SN_{ijn} + \delta S) - \lambda_n(SN_{ijn})}{\delta S} \quad (6\text{-}109)$$

where δS is a preselected, arbitrary increment of saturation. They suggest that δS should be slightly larger in absolute value than the maximum saturation change expected to occur over the time step.

When simultaneous solution methods are used, semi-implicit mobilities can be incorporated into both equations. Sequential methods do not permit this, however, so it is the practice then to use explicit mobilities for the pressure equation and to use the semi-implicit mobilities for the saturation equation alone.

NUMERICAL DISPERSION

It is appropriate, at this point, to remind the reader of the lessons of Chapter 4. The steps we have indicated in this chapter for improving the stability of the difference equations (namely, the use of upstream mobility weighting and the use of semi-implicit mobilities) can be expected to cause some numerical dispersion. Saturation fronts that are supposed to be sharp will probably be smeared to some extent in the numerical calculations. The user must be on guard against this phenomenon.

As was pointed out at the end of Chapter 4, because of the nonlinear nature of multiphase flow problems, there are few quantitative guides concerning the extent of numerical dispersion that may arise in the course of their numerical solution. The user can make some judgment concerning how serious numerical dispersion may be in the multidimensional problem that he wishes to solve by running some one-dimensional problems with the same relative permeability and capillary pressure data over a range of flow rates and for various grid sizes. In multidimensional problems, decreasing the size of the grid blocks will reduce numerical dispersion, but it will also increase the computing cost very substantially. While the user should attempt to assess the degree of numerical dispersion, he may have to accept more of it than he would like, and therefore has to discount accordingly the results that he obtains.

WELL RATES

Injection wells

Specification of individual phase rates of injection normally presents no special problems, since the usual practice is to inject only one phase, such as gas or water. Further, the rates of injection are normally specified as part of the statement of the problem.

Production wells

Production wells present a more complicated situation than do injection wells, in that two phases may be produced, but the relative amounts of each phase produced are not subject to control. Rather, they are functions of saturation in the vicinity of the well and should be subject to the condition that the volumetric rate of production of each phase is proportional to the local mobility of that phase. Thus:

$Q_n/\lambda_n = Q_w/\lambda_w$

Equivalently, we may write:

$Q_n/Q_t = \lambda_n/\lambda_t = f_n$ (6-110)

and:

$Q_w/Q_t = \lambda_w/\lambda_t = f_w$ (6-111)

Explicit production rates

Assuming Q_t is specified in advance, we can frequently use explicit mobilities to calculate the individual phase rates. Thus we have:

$QN_{ijn} = QT_{ijn}FN_{ijn}$

$QW_{ijn} = QT_{ijn}FW_{ijn}$

Semi-implicit production rates

While the above procedure is satisfactory in many cases, there are situations where the use of explicit mobilities to calculate the phase production terms is unsatisfactory, as it leads to wild oscillations in the relative rates of production. Particularly is this true for coning problems, which, we saw earlier, require the use of implicit or semi-implicit mobilities in the flow terms of the difference equations. Exactly the same considerations require that the phase production terms be calculated in an analogous way.

Here, we assume that the total flow rate can be evaluated at the old time level, t_n, while semi-implicit mobilities are used to calculate the individual phase production rates. Thus we have:

$QN_{ijn} = QT_{ijn}[FN_{ijn} + (df_n dS_n)_{ij}(SN_{i,j,n+1} - SN_{ijn})]$

$QW_{ijn} = QT_{ijn}[FW_{ijn} + (df_w/dS_n)_{ij}(SN_{i,j,n+1} - SN_{ijn})]$

A phase production term of this form introduces an additional term containing the unknown saturation, $SN_{i,j,n+1}$ or $SW_{i,j,n+1}$, which is easily incorporated into the saturation equation written for each grid point associated with a production well.

Allocation of total injection or production rates

When a well penetrates only one grid block, the total production rate for the block is equal to that for the well, and the application of the above expressions is straightforward. In the simulation of reservoir cross-sections, however, it is quite common for a single well to penetrate two or more grid blocks. It is also very common in three-dimensional simulations for a single well to penetrate more than one grid block. Then each producing grid block produces a time-varying fraction of the well's total production, and some procedure is required to allocate the well production rate among the several grid blocks each time step.

Ideally, the allocation should take place in such a way that the oil *potential* within the wellbore is the same for each grid block that the well penetrates. Various approaches have been used to approximate this; one of the most common is to allocate the total production according to the total mobility. Thus, if a well penetrates s grid blocks, we might use:

$$\overline{QT}_r = \overline{QT}(\lambda_t)_r \Big/ \sum_{r=1}^{s} (\lambda_t)_r, \quad r = 1, 2, \ldots, s$$

If the grid blocks are not uniform, or if the absolute permeability is not isotropic, then a more complicated version of this equation must be used. Nolen and Berry (1972) discuss this method in considerable detail, as well as another method that considers the flow potentials as well.

Special rate routines

We have, up to this point, taken a rather simplistic view of the problem of setting injection and production rates, and it is beyond the scope of this book to consider more complicated situations. However, the reader should be aware that practical reservoir simulators must involve the interaction among the flow terms at individual grid points, the allocation of total flow for wells which penetrate several grid blocks, as well as the specifications and constraints imposed by the actual operation of the field being simulated. These specifications may involve injection or production rates for one phase, production rates of total liquids, or flowing bottom-hole pressures. They may involve one well or a group of wells. The calculations may need to account for the flowing pressure drop from the bottom of the hole to the wellhead, and perhaps also need to take into account pressure drop in the above-ground flow lines. Constraints may involve limits on water-oil or gas-oil ratios, as well as limits on the bottom-hole pressure. Clearly, a great variety of combinations is possible. It is not uncommon that a user must provide his own rate routine for an otherwise standard reservoir simulator, in order to accommodate special situations unique to the particular field under study. Little has appeared in the literature concerning rate routines; perhaps the most complete discussion is that given by Trimble and McDonald (1976).

NOMENCLATURE

Units

The symbols listed below are defined in terms of SI base units, kg (kilogram), m (meter), and s (second), plus two derived units, N (newton = kg·m/s^2) for force and Pa (pascal = N/m^2 = kg/m·s^2) for pressure.

Symbols

$a = LTX(\Delta t/2\phi S') = LTY(\Delta t/2\phi S')$ [m²]
ANX, ANY = nonwetting-phase transmissibilities [m³/Pa·s]
AWX, AWY = wetting-phase transmissibilities [m³/Pa·s]
$b = LYX(\Delta t/2\phi S') = LDY(\Delta t/2\phi S')$ [m²]
D = depth [m]
f_n = ratio of nonwetting-phase mobility to total mobility, i.e., $\lambda_n/(\lambda_n + \lambda_w)$
f_w = ratio of wetting-phase mobility to total mobility, i.e., $\lambda_w/(\lambda_n + \lambda_w)$
FN = difference solution for function f_n
FW = difference solution for function f_w
FNX, FNY = interval values for function f_n
FWX, FWY = interval values for function f_w
$FWLNX, FWLNY$ = interval values for function $f_w \lambda_n$ [m²/Pa·s]
$i = (-1)^{\frac{1}{2}}$
I = number of grid points or blocks in x-direction
J = number of grid points or blocks in y-direction
k_{rn}, k_{rw} = relative permeability
K = absolute permeability [m²]
KX, KY = interval values of absolute permeabilities, K_x, K_y [m²]
LDX, LDY = difference of interval phase mobilities, equals $(LNX - LWX)$, $(LNY - LWY)$ [m²/Pa·s]
LNX, LNY = interval values of nonwetting-phase mobility [m²/Pa·s]
LTX, LTY = total of interval phase mobilities, equals $(LNX + LWX)$, $(LNY + LWY)$ [m²/Pa·s]
LWX, LWY = interval values of wetting-phase mobility [m²/Pa·s]
p = index in Fourier expansion of error
p_c = capillary pressure [Pa]
p_n, p_w = phase pressure [Pa]
PA = difference solution for average pressure, p_{avg} [Pa]
PC = difference solution for capillary pressure, p_c [Pa]
PN = difference solution for nonwetting-phase pressure, p_n [Pa]
PW = difference solution for wetting-phase pressure, p_w [Pa]
q = index in Fourier expansion of error
q_n, q_w = mass rate of injection (or production, if negative) of a phase per unit volume of reservoir [kg/m³s]
Q_n, Q_w = volumetric rate of injection (or production, if negative) of a phase per unit volume of reservoir [s⁻¹]
QN_{ij}, QW_{ij} = volumetric rate of injection (or production, if negative) of a phase per unit volume of grid block (i, j) [s⁻¹]
$\overline{QN}_{ij}, \overline{QW}_{ij}$ = volumetric rate of injection (or production, if negative) of a phase for grid block (i, j) [m³/s]
S = saturation
$S' = dS_n/dp_c$ [Pa⁻¹]
SN = difference solution for nonwetting-phase saturation, S_n
SW = difference solution for wetting-phase saturation, S_w
SSN, SSW = normalizing factor used in A.D.I. [Pa⁻¹s⁻¹]
$\overline{SSN}, \overline{SSW}$ = normalizing factor used in A.D.I. [m³/Pa·s]
t = time [s]
$\vec{v}$ = flow velocity vector [m/s]
VNX, VNY = interval values of nonwetting-phase flow velocity [m/s]
VTX, VTY = interval values of total flow velocity, equals $(VNX + VWX)$, $(VNY + VWY)$ [m/s]
VWX, VWY = interval values of wetting-phase flow velocity [m/s]
W = distance-weighting parameter for interval mobilities

x = distance [m]
X_p = $(4/\Delta x^2) \sin^2 (p\Delta x/2)$ [m^{-2}]
y = distance [m]
Y_q = $(4/\Delta y^2) \sin^2 (q\Delta y/2)$ [m^{-2}]
α = thickness of reservoir [m]. (α is a general geometric factor defined by eq. (1-13). In this chapter, which deals only with two-dimensional reservoirs, α is equal to thickness.)
β = iteration parameter
γ^n = coefficient in Fourier expansion of error in PA [Pa]
Δx_i = $x_{i+\frac{1}{2}} - x_{i-\frac{1}{2}}$ [m]
Δy_j = $y_{j+\frac{1}{2}} - y_{j-\frac{1}{2}}$ [m]
Δ_x^2 = second difference operator in x-direction, defined by

$$\Delta_x^2 P = (P_{i+1,j} - 2P_{ij} + P_{i-1,j})/\Delta x^2 \ [\text{m}^{-2}]$$

Δ_y^2 = second-difference operator in y-direction, defined by

$$\Delta_y^2 P = (P_{i,j+1} - 2P_{ij} + P_{i,j-1})/\Delta y^2 \ [\text{m}^{-2}]$$

$\Delta(V)$ = difference analog of divergence of velocity, defined by

$$\Delta(V) = [(VX_{i+\frac{1}{2},j} - VX_{i-\frac{1}{2},j})/\Delta x_i] + [(VY_{i,j+\frac{1}{2}} - VY_{i,j-\frac{1}{2}})/\Delta y_j][\text{s}^{-1}]$$

$\Delta_x(L\Delta_x P)$ = second-difference quotient in x-direction, defined by

$$\Delta_x(L\Delta_x P) = [\{LX_{i+\frac{1}{2},j}(P_{i+1,j} - P_{ij})/(x_{i+1} - x_i)\}$$
$$- \{LX_{i-\frac{1}{2},j}(P_{ij} - P_{i-1,j})/(x_i - x_{i-1})\}]/\Delta x_i$$

$\Delta_y(L\Delta_y P)$ = second-difference quotient in y-direction, defined by

$$\Delta_y(L\Delta_y P) = [\{LY_{i,j+\frac{1}{2}}(P_{i,j+1} - P_{ij})/(y_{j+1} - y_j)\}$$
$$- \{LY_{i,j-\frac{1}{2}}(P_{ij} - P_{i,j-1})/(y_j - y_{j-1})\}]/\Delta y_j$$

$\Delta(L\Delta P)$ = combined or two-dimensional second-difference quotient, equal to $\Delta_x(L\Delta_x P) + \Delta_y(L\Delta_y P)$
λ = phase mobility [m^2/Pa · s]
μ = phase viscosity [Pa · s]
ξ^n = coefficient in Fourier expansion of error in PC [Pa]
ρ = phase density [kg/m^3]
ϕ = porosity

Subscripts and superscripts

avg	average of nonwetting- and wetting-phase property
d	difference between nonwetting- and wetting-phase property
i	index in x-direction
j	index in y-direction
k	iteration count
m	undetermined time-step level
n	time step level
n	refers to nonwetting phase
p	index in Fourier expansion of error
q	index in Fourier expansion of error
t	total of nonwetting- and wetting-phase property
w	refers to wetting phase
x	refers to x-direction
y	refers to y-direction

REFERENCES

Ames, W.F., 1969. *Numerical Methods for Partial Differential Equations*. Barnes and Noble, New York, 291 pp.

Blair, P.M. and Weinaug, C.F., 1969. Solution of Two-Phase Flow Problems Using Implicit Difference Equations. *Soc. Pet. Eng. J.*, 9: 417—424; *Trans. Am. Inst. Min. Metall. Pet. Eng.*, 246: 417—424.

Breitenbach, E.A., Thurnau, D.H. and Van Poollen, H.K., 1968. The Fluid Flow Simulation Equations. *Soc. Pet. Eng. Symp. on Numerical Simulation of Reservoir Performance, Dallas, Texas, S.P.E. Paper*, no. 2020: 11 pp.

Brian, P.L.T., 1961. A Finite-Difference Method of High-Order Accuracy for the Solution of Three-Dimensional Transient Heat Conduction Problems. *AIChE (Am. Inst. Chem. Eng.) J.*, 7: 367—370.

Buckley, S.E. and Leverett, M.C., 1942. Mechanism of Fluid Displacement in Sands. *Trans. Am. Inst. Min. Metall. Eng.*, 146: 107—116.

Chaudhari, N.M., 1971. An Improved Numerical Technique for Solving Multi-Dimensional Miscible Displacement Equations. *Soc. Pet. Eng. J.*, 11: 227—284; *Trans. Am. Inst. Min. Metall. Pet. Eng.*, 251: 277—284.

Coats, K.H., Nielsen, R.L., Terhune, M.H. and Weber, A.G., 1967. Simulation of Three-Dimensional, Two-Phase Flow in Oil and Gas Reservoirs. *Soc. Pet. Eng. J.*, 7: 377—388; *Trans. Am. Inst. Min. Metall. Pet. Eng.*, 237: 377—388.

Collins, R.E., 1961. *Flow of Fluids Through Porous Materials*. Reinhold, New York, 270 pp. (Reprinted 1976 by the Petroleum Publishing Company, Tulsa.)

Cook, R.E., Jacoby, R.H. and Ramesh, A.B., 1974. A Beta-Type Reservoir Simulator for Approximating Compositional Effects During Gas Injection. *Soc. Pet. Eng. J.*, 14: 471—481.

Douglas, J., 1961. A Survey of Numerical Methods for Parabolic Differential Equations. In: F.L. Alt (Editor), *Advances in Computers*, vol. 2. Academic Press, New York, pp. 1—54.

Douglas, J., 1962. Alternating-Direction Methods for Three Space Variables. *Numerische Mathematik*, 4: 41—63.

Douglas, J. and Rachford, H.H., 1956. On the Numerical Solution of Heat Conduction Problems in Two and Three Space Variables. *Trans. Am. Math. Soc.*, 82: 421—439.

Douglas, J., Peaceman, D.W. and Rachford, H.H., 1959. A Method for Calculating Multi-Dimensional Immiscible Displacement. *Trans. Am. Inst. Min. Metall. Pet. Eng.*, 216: 297—308.

Dupont, T., Kendall, R.P. and Rachford, H.H., 1968. An Approximate Factorization Procedure for Solving Self-Adjoint Elliptic Difference Equations. *SIAM (Soc. Ind. Appl. Math.) J. Numer. Anal.*, 5: 559—573.

Fagin, R.G. and Stewart, C.H., 1966. A New Approach to the Two-Dimensional Multiphase Reservoir Simulator. *Soc. Pet. Eng. J.*, 6: 175—182; *Trans. Am. Inst. Min. Metall. Pet. Eng.*, 237: 175—182.

Forsythe, G.E. and Wasow, W.R., 1960. *Finite-Difference Methods for Partial Differential Equations*. Wiley, New York, 444 pp.

George, A., 1973. Nested Dissection of a Regular Finite Element Mesh. *SIAM (Soc. Ind. Appl. Math.) J. Numer. Anal.*, 10: 345—363.

Lantz, R.B., 1971. Quantitative Evaluation of Numerical Diffusion (Truncation Error). *Soc. Pet. Eng. J.*, 11: 315—320; *Trans. Am. Inst. Min. Metall. Pet. Eng.*, 251: 315—320.

Laumbach, D.D., 1975, A High-Accuracy, Finite-Difference Technique for Treating the Convection-Diffusion Equation. *Soc. Pet. Eng. J.*, 15: 517—531.

Letkeman, J.P. and Ridings, R.L., 1970. A Numerical Coning Model. *Soc. Pet. Eng. J.*, 10: 418—424; *Trans. Am. Inst. Min. Metall. Pet. Eng.*, 249: 418—424.

MacDonald, R.C. and Coats, K.H., 1970. Methods for Numerical Simulation of Water and Gas Coning. *Soc. Pet. Eng. J.*, 10: 425—436; *Trans. Am. Inst. Min. Metall. Pet. Eng.*, 249: 425—436.

Nolen, J.S. and Berry, D.W., 1972. Tests of the Stability and Time-Step Sensitivity of Semi-Implicit Reservoir Simulation Techniques. *Soc. Pet. Eng. J.*, 12: 253—266; *Trans. Am. Inst. Min. Metall. Pet. Eng.*, 253: 253—266.

Peaceman, D.W. and Rachford, H.H., 1955. The Numerical Solution of Parabolic and Elliptic Differential Equations. *SIAM (Soc. Ind. Appl. Math.) J.*, 3: 28—41.

Pearcy, C., 1962. On Convergence of Alternating-Direction Procedures. *Numerische Mathematik*, 4: 172—176.

Peery, J.H. and Herron, E.H., 1969. Three-Phase Reservoir Simulation. *J. Pet. Technol.*, 21: 211—220; *Trans. Am. Inst. Min. Metall. Pet. Eng.*, 246: 211—220.

Price, H.S. and Coats, K.H., 1974. Direct Methods in Reservoir Simulation. *Soc. Pet. Eng. J.*, 14: 295—308; *Trans. Am. Inst. Min. Metall. Pet. Eng.*, 257: 295—308.

Price, H.S., Cavendish, J.C. and Varga, R.S., 1968. Numerical Methods of Higher-Order Accuracy for Diffusion-Convection Equations. *Soc. Pet. Eng. J.*, 8: 293—303; *Trans. Am. Inst. Min. Metall. Pet. Eng.*, 243: 293—303.

Richtmyer, R.D., 1957. *Difference Methods for Initial-Value Problems*. Interscience, New York, 238 pp.

Settari, A. and Aziz, K., 1973. A Generalization of the Additive Correction Methods for the Iterative Solution of Matrix Equations. *SIAM (Soc. Ind. Appl. Math.) J. Numer. Anal.*, 10: 506—521.

Settari, A. and Aziz, K., 1974. A Computer Model for Two-Phase Coning Simulation. *Soc. Pet. Eng. J.*, 14: 221—236.

Sheffield, M., 1969. Three-Phase Fluid Flow Including Gravitational, Viscous and Capillary Forces. *Soc. Pet. Eng. J.*, 9: 232—246; *Trans. Am. Inst. Min. Metall. Pet. Eng.*, 246: 232—246.

Snyder, L.J., 1969. Two-Phase Reservoir Flow Calculations. *Soc. Pet. Eng. J.*, 9: 170—182.

Sonier, F., Besset, P. and Ombret, O., 1973. A Numerical Model of Multiphase Flow Around a Well. *Soc. Pet. Eng. J.*, 13: 311—320.

Spillette, A.G., Hillestad, J.G. and Stone, H.L., 1973. A High-Stability Sequential Solution Approach to Reservoir Simulation. *Soc. Pet. Eng. 48th Ann. Meet., Las Vegas, Nev., S.P.E. Paper*, no. 4542, 14 pp.

Stone, H.L., 1968. Iterative Solution of Implicit Approximations of Multidimensional Partial Differential Equations. *SIAM (Soc. Ind. Appl. Math.) J. Numer. Anal.*, 5: 530—558.

Stone, H.L. and Brian, P.L.T., 1963. Numerical Solution of Convective Transport Problems. *AIChE (Am. Inst. Chem. Eng.) J.*, 9: 681—688.

Todd, M.R., O'Dell, P.M. and Hirasaki, G.J., 1972. Methods for Increased Accuracy in Numerical Reservoir Simulators. *Soc. Pet. Eng. J.*, 12: 515—530; *Trans. Am. Inst. Min. Metall. Pet. Eng.*, 253: 515—530.

Trimble, R.H. and McDonald, A.E., 1976. A Strongly Coupled, Implicit Well Coning Model. *Soc. Pet. Eng. 4th Symp. on Numerical Simulation of Reservoir Performance, Los Angeles, Cal., S.P.E. Paper*, no. 5738: 11 pp.

Varga, R.S., 1962. *Matrix Iterative Analysis*. Prentice-Hall, Englewood Cliffs, N.J., 322 pp.

Watts, J.W., 1971. An Iterative Matrix Solution Method Suitable for Anisotropic Problems. *Soc. Pet. Eng. J.*, 11: 47—51; *Trans. Am. Inst. Min. Metall. Pet. Eng.*, 251: 47—51.

Watts, J.W., 1973. A Method for Improving Line Successive Overrelaxation in Anisotropic Problems — A Theoretical Analysis. *Soc. Pet. Eng. J.*, 13: 105—118; *Trans. Am. Inst. Min. Metall. Pet. Eng.*, 255: 105—118.

Weinstein, H.G., Stone, H.L. and Kwan, T.V., 1969. An Iterative Procedure for Solution of Systems of Parabolic and Elliptic Equations in Three Dimensions. *Ind. Eng. Chem. Fund.*, 8: 281—287.

Weinstein, H.G., Stone, H.L. and Kwan, T.V., 1970. Simultaneous Solution of Multiphase Reservoir Flow Equations. *Soc. Pet. Eng. J.*, 10: 99—110.

Young, D.M., 1962. The Numerical Solution of Elliptic and Parabolic Partial Differential Equations. In: J. Todd (Editor), *Survey of Numerical Analysis*. McGraw-Hill, New York, pp. 380—438.

Young, D.M., 1971. *Iterative Solution of Large Linear Systems*. Academic Press, New York, 570 pp.

INDEX

Accumulation, rate of, 3—6, 16, 152
Accuracy, *see* Truncation error
Additive corrections, method of, 115—119
Allocation of well rates, 165, 166
Alternating-direction implicit procedure, 57—63
— — — —, for 2-phase flow, 144—146
— — — —, three-dimensional, 60—63
Alternating-direction iteration, 120—127
— — —, for 2-phase flow, 148—150
Alternative equations for two-phase flow, 17—20, 140—141, 153, 154
Amplification factor, *see also* Stability, 48
Anisotropic permeability, 83, 143, 166
Anisotropy, effect on convergence, 100, 112, 114, 115, 117, 120
Approximate factorization, 128—134
Area, cross-sectional, 3, 7
Asymptotic behavior of error, 43
Auxiliary relations, 8, 25—27
Average pressure, 17, 140, 144, 154, 159

Backward-difference equation, 49, 56
Backward-difference quotient, 36
Band matrix, 86
— — algorithm, 87—90
— — —, use of, 90, 150, 151
Bitridiagonal equations, 149
Black-oil model, 27—30
Blair and Weinaug, 161
Block-centered grid, 38—39, 40
Block-tridiagonal matrix, 104, 105
Boundary conditions, 8, 9, 41
— —, Dirichlet, 93, 97, 100, 112, 115
— —, Neumann, 83, 97, 98, 112, 113, 122
— —, no-flow, 8, 9, 41
— —, reflection, 41
Brian method, 62, 63
Buckley-Leverett equation, 22, 141, 157

Capillary pressure, 14, 15, 18, 21, 23, 27, 140, 144, 154
— — term, treatment of, 157, 159
Centered-difference quotient, 36, 37
Characteristics, method of, 22
Checkerboard ordering, 107

Chord slope, 163
Classification of differential equations, 12—14
Coefficient matrix, 85—87, 92, 95, 128
Commutivity, 121
Component balance, 25
Compositional model, 24—27
— —, limited, 30—32
Compressibility, 9, 10, 17, 18, 139, 146
—, single-phase, 9—11
—, two-phase, 16, 17, 18
Computing rectangle, 85
Concentration, 21
Conditional stability
— —, forward-difference equation, 49, 56
— —, hyperbolic difference equations, 69, 70, 71
— —, using explicit mobilities, 160—161
Coning problems, 118, 160
Conservation of mass, 4, 6, 7, 16, 23, 25, 152
Continuity, *see* Conservation of mass
Convection equation, 14, 21, 65
Convergence
—, A.D.I., 121—126
—, Jacobi iteration, 97—103, 110, 111, 119
—, line-Jacobi iteration, 113—115
—, line relaxation, 118, 119
—, relation to eigen values, 101
—, S.I.P., 133—134
—, SOR, 103—112, 119
Crank-Nicolson difference equation, 52, 57

Darcy's law
— —, single-phase, 2, 3, 7
— —, 2-phase, 15
— —, 3-phase, 23
Density, 3, 8, 26, 28, 30, 31
Depth, 3
Difference equation
— —, backward, 49, 50, 56
— —, Crank-Nicolson, 52, 57
— —, forward, 41, 45—49, 55
— —, hyperbolic, 65—81

— —, —, special cases, 67
— —, three-level, 53
— —, time-centered, explicit, 53
Difference operators, 37, 38, 97, 141, 142, 159
— —, for velocity term, 158
Differential equations
— —, classification of, 12—14
— —, derivations, 2—32
— — — for single-phase flow, 2—12
— — — for 2-phase flow, 14—22, 139—141
— — — for 3-phase flow, 23—24, 25, 29—30, 32
Differential operators, 6, 7
Differential volume elements, 4
Diffusion-convection equation, 21, 74—81
Diffusivity, 12, 21, 74
Direct solution, band matrix, 87—90
Dirichlet boundary conditions, 93, 97, 100, 112, 115
— — —, error expansion for, 97
Dispersion, numerical, 74—81, 164
Distance-weighting, 65, 143
Divergence operator, 6
Douglas method, 62, 63
Douglas-Rachford iteration, 127
Douglas-Rachford method, 60—62
Downstream weighting, 66
Dufort-Frankel, 53—55
Dupont, Kendall and Rachford, 132, 133

Eigenvalue
—, definition of, 101
— of Jacobi iteration matrix, 103, 107
— of SOR iteration matrix, 104—110
Elliptic equations, 13, 18, 83—134
Equilibrium, phase, 27
Error equation
— —, A.D.I., 122
— —, forward-difference equation, 46, 47
— —, Jacobi iteration, 96
— —, line-Jacobi, 113
— —, SOR, 96
Error expansions, 48, 56, 97, 98, 102
— —, Dirichlet boundary conditions, 97
— —, Neumann boundary conditions, 97, 98
Error function, 76
Error in iteration, 95, 101
Error, truncation, see Truncation error
Explicit difference equation, 45, 46, 47, 53, 55, 66, 144
Explicit mobilities, 160—161

Factorization
—, approximate, 128—134
—, direct, 87—90
Finite differences, 35—42
First-difference quotient, 35—37
Formation volume factors, 28—31
Forward-difference equation, 41, 45—49, 55
Forward-difference quotient, 36
Fourier equation, see Heat conduction equation
Fractional flow function, 22
Fractional mobility function, 19, 159

Gas, ideal, 11
Gas percolation, 160
Gas solubility in oil, 28, 31
Gas solubility in water, 31
Gauss-Seidel, see Successive displacements
Geometric factor function, 7, 139
Global truncation error, 42
Gradient operator, 7
Gravity, 3, 139
Grid spacing, variable, 38, 40, 83, 166
Grid systems, 38—41

Harmonic analysis of convergence, 97—100, 121—123
Harmonic analysis of stability, 46, 48—50, 52—54, 56, 57, 58—61, 68—71, 145, 147, 154
Heat conduction equation, 10, 12, 21, 41, 45ff
Hyperbolic difference equations, 65—81
— — —, special cases, 67
Hyperbolic equations, 13, 21, 65—81

Ideal gas, 11
Ideal liquid, 9—10
Implicit difference equations, 49, 52, 56, 57, 66
Implicit mobilities, 161
Implicit procedure for 2-phase flow, simultaneous, 146—153
Incompressible flow, 12, 83, 139
Indexing of grid points, 38
Indexing of matrix elements, 85
Injection rates, 3, 5, 6, 8, 17, 84, 140, 142, 164, 165, 166
Interval mobilities, 143
Interval permeabilities, 143
Interval velocities, 158
Iterative methods for elliptic problems, 91—134

Jacobi iteration, 94, 95, 96
— —, convergence, 97—103, 110, 111, 119

Laplace's equation, 12, 13, 93, 111
Laplacian operator, 7
Leapfrog method, 154—155
Line-Jacobi iteration, 112—115, 119
Line relaxation, 112—120
Local truncation error, definition, 42

Mass, conservation of, 4, 6, 7, 16, 23, 25, 152
Mass transfer between phases, 24—32, 139
Material balance, 151, 152—153
Matrix, band, 86, 87—90
Matrix notation, 84, 85, 96
Midpoint weighting, 66, 143
Miscible displacement, 21
Mobilities
—, explicit, 160—161
—, implicit, 161
—, interval, 143
—, phase, 18, 140
—, semi-implicit, 162—164
Model, reservoir, 1

Nested dissection, 91
Neumann boundary conditions, 83, 97, 98, 112, 113, 122
— — —, error expansion for, 97, 98
Newtonian iteration, 162
No-flow boundary conditions, 8, 9, 83
Nonlinear coefficient, S', 151
Nonstandard ordering, 91
Nonuniform grid spacing, 38, 40, 83, 166
Normalizing factor for A.D.I., 120, 149
Numerical dispersion, 74—81, 164
Numerical model, 1

One-dimensional flow, 3, 22
Optimum parameter, SOR, 109—110, 111
Optimum parameters for A.D.I., 124—126
Ordering of coefficient matrix, 85, 90—91, 104, 107
Order of error, see also Truncation error, 36
Overrelaxation, see also Successive overrelaxation, 93
Overshoot, 78, 79

Parabolic equations, 12, 17, 21, 37, 45—63
Parameters for A.D.I., 121—127, 150
Parameters for S.I.P., 133—134

Peaceman-Rachford implicit procedure, 57—60
Peaceman-Rachford iteration procedure, 120—127
Permeability
—, absolute, 2, 143
—, anisotropic, 83, 143, 166
—, interval, 143
—, relative, 14, 15, 26
Phase equilibrium, 27
Phase mobilities, 18, 140
Point-centered grid, 39—40, 41
Poisson's equation, 12, 13
Porosity, 3, 6
Potential, 12, 166
Pressure, average, see Average pressure
Pressure, capillary, see Capillary pressure
Pressure differential equation, 17—18, 24, 141
Production rate, 3, 5, 6, 8, 17, 84, 140, 142, 164—166
Property A, 104

Rachford, see Douglas-Rachford, Dupont-Kendall-Rachford, Peaceman-Rachford
Rate routines, 166
Rectangle, computing, 85
Reflection boundary condition, 41
Relative permeability, 14, 15, 26
Relaxation methods, 92—120
Reservoir simulation, definition, 1
Residual, 93, 115, 116, 130

Saturation creep, 155—156, 157, 160
Saturation, definition of, 14
Saturation differential equations, 19—22, 24, 141, 159
Second-difference operator, 37, 38, 56, 97, 141—142
Second-difference quotient, 37—38, 41, 56, 83, 141—142
Semi-implicit
— — capillary pressure, 159
— — mobilities, 162—164
— — production rate, 165
Sequential methods for two-phase flow, 153—160
Settari and Aziz, 2-D method of additive corrections, 118
Simultaneous displacements, see Jacobi iteration
Simultaneous implicit procedure for two-phase flow, 146—153

Single-phase flow, 2—12
— — —, special cases, 9—12
S.I.P. (Strongly implicit procedure), 128, 132—134
— for multiphase flow, 150
—, 3-dimensional, 134
Solubility of gas in oil, 28, 31
Solubility of gas in water, 31
SOR, *see* Successive overrelaxation
Southwell relaxation, 93—94
Spillette, Hillestad and Stone, 157—160
Stability, 46ff
—, alternating-direction implicit procedure, 58—61
—, — — — — for 2-phase flow, 145—146
—, backward-difference equation, 49, 50, 56
—, conditional, *see* conditional stability
—, Crank-Nicolson, 52, 57
—, Douglas-Rachford, 61
—, Dufort-Frankel, 54
—, forward-difference equation, 46—49, 56
—, hyperbolic difference equations, 68—71, 80
—, leapfrog method, 154
—, Peaceman-Rachford, 58, 60
—, simultaneous implicit procedure for 2-phase flow, 147
Stone's S.I.P. iteration, 128, 132—134
Strongly implicit procedure, 128, 132—134
Successive displacements, 94, 108, 111
Successive overrelaxation, 94, 96, 103—112
— —, convergence, 103—112, 119
— —, optimum parameter, 109—110, 111
Surface tension, 14

Teylor series, 35, 36, 37, 54
Thickness, 5, 7
Three-dimensional alternating-direction, 60—63
Three-dimensional flow, 6
Three-dimensional S.I.P., 134
Three-phase flow, 23ff
Time-centered explicit difference equation, 53
Time-step restriction, *see* Conditional stability
Time-weighting, 66, 161
Total velocity
— —, sequential method using, 157—160

— —, 2-phase flow, 18, 19, 141, 158
— —, 3-phase flow, 24
Transmissibility, 84, 149
Tridiagonal algorithm, 50—52
— —, use of, 50, 52, 58, 76, 112, 116, 120
Truncation error, 36, 37, 41—43
— —, asymptotic behavior, 43
— —, backward-difference equation, 53, 57
— —, Crank-Nicolson, 53, 57
— —, Douglas-Rachford, 61
— —, Dufort-Frankel, 54—55
— —, forward-difference equation, 52, 57
— —, global, 42
— —, hyperbolic difference equations, 71—75
— —, local, 42
— —, multiphase flow, 151, 164
— —, Peaceman-Rachford, 59—60
Two-dimensional flow, 5
Two-phase flow, 14—22, 139—166

Upstream weighting, 66, 143

Variable grid spacing, 38, 40, 83, 166
Velocity
—, difficulties due to high, 160
—, phase, 15, 19, 23, 141
—, —, interval, 158
—, single-phase, 2, 3, 7
—, total, 18—19, 141
—, —, interval, 158
Viscosity, 2, 26
Volatility, 30, 31
Volume, differential elements of, 4
Von Neumann criterion for stability, 48
Von Neumann stability analysis, *see* Harmonic analysis of stability

Wachspress parameters for A.D.I., 125
Watts method of additive corrections, 115—118
Wave equation, 13
Weighting
—, distance, 65, 143
—, midpoint, 66, 143
—, time, 66, 161
—, upstream, 66, 143
Well rates, 164—166
Work for factorization, 90, 91
Work for iteration, 92

Young's rate of convergence, 98, 102, 109—111, 119, 126